CHILDHOOD OBESITY

CONTEMPORARY ISSUES

Society for the Study of Human Biology Series

43 The Changing Face of Disease: Implications for Society
Edited by C. G. N. Mascie-Taylor, J. Peters and S. T. McGarvey

44 Childhood Obesity: Contemporary Issues
Edited by Noel Cameron, Nicholas G. Norgan, and George T. H. Ellison

Numbers 1–9 were published by Pergamon Press, Headington Hill Hall, Headington, Oxford OX3 0BY. Numbers 10–24 were published by Taylor & Francis Ltd, 10–14 Macklin Street, London WC2B 5NF. Numbers 25–40 were published by Cambridge University Press, The Pitt Building, Trumpington Street Cambridge CB2 1RP. Further details and prices of back-list numbers are available from the Secretary of the Society for the Study of Human Biology.

CHILDHOOD OBESITY

CONTEMPORARY ISSUES

Edited by

**N. Cameron, N. G. Norgan
and G. T. H. Ellison**

618
.92398
S989c

12/05
Taylor & Francis
Taylor & Francis Group

Boca Raton   London   New York   Singapore

A CRC title, part of the Taylor & Francis imprint, a member of the
Taylor & Francis Group, the academic division of T&F Informa plc.

Published in 2006 by
CRC Press
Taylor & Francis Group
6000 Broken Sound Parkway NW, Suite 300
Boca Raton, FL 33487-2742

Library of Congress Cataloging-in-Publication Data

Childhood obesity : contemporary issues / [edited] by Noël Cameron, Nicholas G. Norgan, and George
T.H. Ellison.
    p. cm. – (Symposia of the Society for the Study of Human Biology ; no. 45)
    ISBN 0-8493-2857-8 (alk. paper)
    1. Obesity in children—Congresses. I. Cameron. Noël. II. Norgan, Nicholas G. III. Ellison, George
T.H. IV. Series.

RJ399.C6C4753 2005
618.92'398—dc22   2005041938

Taylor & Francis Group
is the Academic Division of T&F Informa plc.

Visit the Taylor & Francis Web site at
http://www.taylorandfrancis.com

and the CRC Press Web site at
http://www.crcpress.com

CONTENTS

Part I: ASSESSING THE SCOPE AND IMPACT OF CHILDHOOD OBESITY

LIST OF ILLUSTRATIONS

Figure 1.1   Body Mass Index Centiles for British Boys. The Centile Curves are Spaced Two Thirds of an SD Score Apart. Also Shown are Body Mass Index Values of 25 and 30 kg/m² at Age 18, With Extra Centile Curves Drawn Through Them.

Figure 1.2   Trends Over Time in the Prevalence of Overweight in Boys From 10 Countries, Based on the IOTF International Cutoffs (Cole et al., 2000).

Figure 2.1   Obesity Patterns Across the Developing World. (From Popkin, B.M., *Public Health Nutr.*, 5, 93–103, 2002a).

Figure 2.2   Obesity Trends Among Adults in Selected Developing World Countries (The Annual Percentage Point Increase in Prevalence). (From Popkin, B.M., *Public Health Nutr.*, 5, 93–103, 2002a).

Figure 2.3   Relationship Between Education and Maternal Obesity for Women Aged 19 to 49 Years, Demographic and Health Surveys. Estimated Using a Two-Level Logistic Model, Including Explanatory Variables: Age, SES, Log Country GNP, and Interaction Between SES and Country GNP. Curves Show Predicted Obesity Estimates Ranging From Low Education (Quartile 1) to High Education (Quartile 4). (From Monteiro, C.A., et al., *Lancet*, in Press).

Figure 2.4   Relationship Between Household Income Per Capita and the Prevalence of Overweight (Defined Using IOTF Cutoff Points). Prevalence % are in Boxes, and T-bars Show Standard Errors. Income Tertiles were Used in Each Country to Indicate Low, Medium, and High Levels. For the First and Second Surveys, Respectively, $n = 56,295$ and 4875 for Brazil, 3014 and 2688 for China, 6883 and 2152 for Russia, and 4472 and 6108 for the U.S. (From Wang, Y., et al., *Am. J. Clin. Nutr.*, 75, 971–977, 2002).

Figure 2.5   Overweight Prevalence by Category of Income in U.S. for White, Black, Hispanic, and Asian Adolescents, Using Nationally Representative Data From the National Longitudinal Study of Adolescent Health; $n = 13, 113$. *Statistically Significant Differences From Same-Sex, Non-Hispanic Whites, $p < .05$. **Statistically Significant Differences From Same-Sex, Non-Hispanic Whites, $p < .01$. (From Gordon-Larsen, P., et al., *Obes. Res.*, 11, 121–129, 2003).

Figure 2.6 Shifts in Obesity From Adolescence (1995) to Young Adulthood (2001) in White, Black, Hispanic, and Asian U.S. Adolescents and Adults Using Nationally Representative, Longitudinal Data From the National Longitudinal Study of Adolescent Health; $n = 9795$. *Statistically Significant Differences From Same-Sex, Non-Hispanic Whites, p # .003. (From Gordon-Larsen, P., et al., *Am. J. Clin. Nutr.*, 80, 569–575, 2004).

Figure 4.1 Self-Esteem in 12-Year-Old Children Not Victimized (Open Columns), Fat-Teased Boys (Closed Columns), and Fat-Teased Girls (Hatched Columns). [Data From Hill and Murphy, *Int. J. Obes.*, 24 (Suppl. 1), 161, 2000.] Significant Impact of Victimization, *p < .05 and **p < .001.

Figure 4.2 Self-Esteem in Adolescents Attending a Summer Weight-Loss Camp at the Start (Dark Columns) and End (Light Columns) of their Stay. [Data from Walker et al., *Int. J. Obes.*, 27, 748–754, 2003.] Significant Pre-post Differences, **p < .001.

Figure 6.1 Body Shape silhouettes of increasing obesity (after stunkard et al., 1983).

Figure 7.1 The Proportion of Child Questionnaires Returned From Birth to Age 9 Years in the Avon Longitudinal Study of Parents and Children (ALSPAC). [From The Avon Longitudinal Study of Parents and Children (ALSPAC)].

Figure 9.1 School Day Pattern of Moderate to Vigorous Physical Activity (MVPA) in English Primary School Children.

Figure 9.2 A: School Day Pattern of Physical Activity in Obese (>95th Centile for BMI) vs. Nonobese (<95th Centile for BMI) Primary School-Aged Boys. B: School Day Pattern of Physical Activity in Obese (>95th Centile for BMI) vs. Nonobese (<95th Centile for BMI) Primary School-Aged Girls.

Figure 9.3 A: Weekend Pattern of Physical Activity in Obese (>95th Centile for BMI) vs. Nonobese (<95th Centile for BMI) Primary School-Aged Boys. B: Weekend Pattern of Physical Activity in Obese (>95th Centile for BMI) vs. Nonobese (<95th Centile for BMI) Primary School-Aged Girls.

Figure 9.4 A: Distribution of Physical Activity Level by BMI Centile for Boys (10.5 Years). B: Distribution of Physical Activity Level by BMI Centile for Girls (10.5 Years).

Figure 9.5 Level of Moderate to Vigorous Physical Activity (MVPA) of Boys Who Walk to School Compared With Those Traveling to School by Car.

Figure 9.6 Time Spent in Different Activities Between 3:00 and 8:00 P.M. in Boys, Classified by Mode of Travel to School.

Figure 10.1 A Schematic Diagram Outlining the Biological Mismatch Between the Rate (Low) of Energy Expenditure (EE; Exercise) Compared With the Rate (High) of Energy Intake (EI; Eating).

Figure 10.2 Mean Energy Intake and Energy Expenditure (Exp) (in MJ) During Days of Imposed Activity (Act) and Sedentariness (Sed) in Subjects That ate. Low-Fat (LF) or High-Fat (HF) Diets. Energy Intake is not Downregulated in

LIST OF TABLES

CONTRIBUTORS

P. Betts
Southampton University Hospitals Trust
Southampton, United Kingdom

J. E. Blundell
BioPsychology Group
School of Psychology
University of Leeds
Leeds, United Kingdom

G. Brunton
Social Science Research Unit
Institute of Education
London, United Kingdom

E. Bryant
BioPsychology Group
School of Psychology
University of Leeds
Leeds, United Kingdom

N. Cameron
Loughborough University
Dept of Human Sciences
Loughborough, United Kingdom

T. J. Cole
Centre for Paediatric Epidemiology and
   Biostatistics
Institute of Child Health
University College London
London, United Kingdom

A. R. Cooper
Department of Exercise and Health Sciences
Centre for Sport, Exercise, and Health
University of Bristol
Bristol, United Kingdom

D. B. Dunger
Department of Paediatrics, Addenbrooke's
   Hospital
University of Cambridge
Cambridge, United Kingdom

L. Edmunds
Childhood Obesity Clinic
Bristol Royal Hospital for Children
Bristol, United Kingdom

G. T. H. Ellison
St. George's Hospital Medical School
University of London
London, United Kingdom

L. Foster
Centre for Paediatric Epidemiology and
   Biostatistics
Institute of Child Health
London, United Kingdom

P. Gordon-Larsen
Carolina Population Center
University of North Carolina
Chapel Hill, North Carolina

A. Harden
Social Science Research Unit
Institute of Education
London, United Kingdom

G. Hastings
Department of Marketing
Institute of Social Marketing
University of Stirling
Stirling, United Kingdom

A. Hill
Academic Unit of Psychiatry and Behavioural
Sciences
Leeds University School of Medicine
Leeds, United Kingdom

J. Kavanagh
Social Science Research Unit
Institute of Education
London, United Kingdom

N. A. King
BioPsychology Group
School of Psychology
University of Leeds
Leeds, United Kingdom

M. B. E. Livingstone
Northern Ireland Centre for Diet and Health
University of Ulster
Coleraine, Co. Londonderry, United Kingdom

L. McDermott
Department of Marketing
Institute of Social Marketing
University of Stirling
Stirling, United Kingdom

N. G. Norgan
Loughborough University
Dept of Humen Sciences
Loughborough, United Kingdom

A. Ness
The Avon Longitudinal Study of Parents
and Children
Unit of Paediatric and Perinatal
Epidemiology
Department of Community-Based
Medicine
University of Bristol
Bristol, United Kingdom

A. Oakley
Social Science Research Unit
Institute of Education
London, United Kingdom

S. Oliver
Social Science Research Unit
Institute of Education
London, United Kingdom

K. K. Ong
Department of Paediatrics, Addenbrooke's
Hospital
University of Cambridge
Cambridge, United Kingdom

A. Page
Department of Exercise and Health Sciences
Centre for Sport, Exercise, and Health
University of Bristol
Bristol, United Kingdom

T. Parsons
Paediatric Epidemiology and Biostatistics
Institute of Child Health
University College London
London, United Kingdom

B. M. Popkin
Carolina Population Center
School of Public Health
University of North Carolina
Chapel Hill, North Carolina

R. Rees
Social Science Research Unit
Institute of Education
London, United Kingdom

F. Regan
Department of Paediatrics, Addenbrooke's
  Hospital
University of Cambridge
Cambridge, United Kingdom

J. J. Reilly
University of Glasgow Division of Develop-
  mental Medicine
Queen Mother's Hospital
Glasgow, Scotland, United Kingdom

K. L. Rennie
MRC Human Nutrition Research
Cambridge, United Kingdom

M. J. Rudolf
Community Paediatrics
University of Leeds
Leeds, United Kingdom

M. Stead
Department of Marketing
Institute of Social Marketing
University of Stirling
Stirling, United Kingdom

K. Sutcliffe
Social Science Research Unit
Institute of Education
London, United Kingdom

J. Thomas
Social Science Research Unit
Institute of Education
London, United Kingdom

INTRODUCTION

Noël Cameron, Nicholas G. Norgan,
and George T.H. Ellison

In December 2003, Loughborough University hosted a Symposium on
Childhood Obesity under the auspices of the Society for the Study of Human
Biology (SSHB) and the BioSocial Society (BSS). The theme was thought
to be a critical issue, not only to biologists and biosocial scientists, but to
all those who deal, professionally and academically, with children. Indeed,
it could be said that the theme was critical to all those who care about the
future health and well being of children. The organizers had no qualms
about making the central issues parochial and concentrating on the problem
of childhood obesity in the U.K. In the U.S., 40% of children are now either
obese or in danger of becoming obese. The costs of this epidemic to the
U.S. are staggering, accounting for billions of dollars in health care alone.
It is estimated that, unless the U.K. vigorously tackles its current pattern of
childhood obesity, it will match the U.S. figures within 10 years. The National
Audit Office (NAO) estimates that obesity is costing the U.K. National Health
Service £500 million per year and the cost to society in lost work time and
economic output is estimated at £2 billion per year. By 2010, it is thought
the costs will spiral, and the total direct and indirect costs to the NHS and
wider economy will be £3.6 billion (NAO, 2001).

As staggering as these estimates are, they hide the human cost of obesity.
In the U.K. 20% of children are currently overweight or obese, but this
figure differs by sex and age such that higher percentages of adolescent
girls face the consequences of obesity. The most recent Health Survey for
England (HSE) figures (2001) identify 8.5% of 6-year-old children and 15%
of 15-year-old adolescents as obese. The 2004 Government White Paper

on the Health of the Nation identifies 17% of children aged 2 to 15 years as obese. Obese children are either sick now or will be in the future. Obese children are at significantly greater risk for Types 1 and 2 diabetes, as well as debilitating circulatory, respiratory, and skeletal problems. In addition, psychosocial problems, low self-esteem, and low self-image are the most common and immediate forms of morbidity associated with childhood obesity. These problems are pervasive in the families of obese children and increase as the child gets older. Obese children invariably become obese adults. Obese adults face premature disability and early death. The NAO estimated that in 1998 obesity caused more than 250,000 cases of Type 2 diabetes, 28,000 heart attacks, and 750,000 cases of hypertension (Chief Medical Officer's Annual Report 2002 DoH, 2003; NAO, 2001).

Obesity is the direct result of ingesting too much energy in the form of food and expending too little energy in the form of physical activity. The current increase in the prevalence of obesity has little to do with genetic factors but much to do with changes in the way in which children and adults are eating and exercising. There has been a well-documented shift away from traditional high-fiber, low-fat diets toward high-energy, high-fat, low-fiber diets and an increase in sedentary behaviors. Childhood obesity in particular cannot solely be blamed on increased television viewing and use of computer games (Marshall et al., 2004), but clearly a lifestyle change has emerged: children undertake less organized physical activity and travel to and from school via motorized transport rather than on foot or by bicycle.

These factors do not affect all parts of the population equally. Unequivocal evidence highlights a significant socioeconomic influence that results in obesity being greatest amongst the socioeconomically disadvantaged. High-energy, high-fat foods are extensively marketed and are inexpensive. Structured physical activity outside the school requires membership or entrance fees that further strain the low-income homes characterized by large families.

The epidemic of obesity that threatens to affect 40% of the population in the next 20 years must be dealt with. To do so with the greatest chance of success, it is necessary to arm ourselves with the most extensive knowledge about the causes, and consequences, of childhood obesity. Prevent obesity in childhood and adolescence and not only do you improve their life experience but you also prevent a significant amount of adult obesity.

The Loughborough symposium brought together acknowledged experts from throughout the U.K. and significant researchers from the U.S. Their brief was simple: talk about the causes and consequences of childhood obesity. We have organized this volume of proceedings in three sections; the scope and impact of childhood obesity, the role of biological and social processes in the etiology of childhood obesity, and the prevention and treatment of childhood obesity.

THE SCOPE AND IMPACT OF CHILDHOOD OBESITY

For a disease with such all encompassing and serious consequences, its diagnosis is remarkably easy not just on a national but on a global basis. The global indicator of choice is body-mass index (BMI), the ratio of weight in kilograms to height in meters squared [Wt (kg)/Ht^2 (m)]. In adults, the cut-off points are universally agreed at 25 for overweight and 30 for obesity. In childhood, agreeing on appropriate cut-offs is more complicated because children do not reach BMIs of 25 and 30 until they are fully mature. Thus one needs to have a cut-off curve for BMI that reflects appropriate childhood BMI levels but is still effective in identifying childhood overweight and obesity. Europe has agreed that it will use the International Obesity Task Force charts specifically produced for this purpose (Chapter 1), which identify the 91st and 98th BMI centiles as being indicative of overweight and obesity in childhood. The U.S. has decided to use slightly different criteria based on its own nationally representative BMI growth charts (Chapter 2). If a child has a BMI greater than the 85th centile, he or she is classified as being "in danger of obesity" and at the 95th centile the child is classified as obese. It is perhaps disappointing that there cannot be an agreed universal diagnostic definition of obesity; however, as long as effective action follows identification, the results of different standards may be inconsequential.

The consequences of childhood obesity are serious and far reaching, with both physical and psychological components in the complexity of possible treatments. Only recently has the full picture of the consequences of childhood obesity been emerging from the global evidence. Type 2 (non-insulin-dependent) diabetes was thought to be the only adult consequence now evident in children; however, recently, Type 1 diabetes has also been identified. Cardiovascular risks factors such as centralized fat, hypertension, increased triglycerides, high fibrinogen and insulin levels, increased left ventricular mass, rapid weight changes, and clustering of risk factors are now in evidence in children. During puberty, the emergence of respiratory, orthopedic, gastrointestinal, and psychological problems is becoming commonplace, with psycho-social problems especially problematic among adolescent girls (Chapter 3).

The psycho-social consequences of obesity are complex and highlight the difficulty of seeking effective treatment through behavioral change (Chapter 4). Although reduced self-esteem is universally recognized as a result of overweight and obesity, its importance to the individual differs by gender and age. Children and adolescent boys appear to be less affected, but adolescent girls suffer the most. In particular, overweight or obese preteen children do not appear to experience discrimination by their peers.

However, this effect declines with increasing age, with obese girls particularly prone to being assessed as unattractive. Experience with children attending a "weight loss camp" suggests that it might be possible to improve self-esteem through weight loss, but there remains little evidence of a causal link between the two, and no evidence that discounting social domains, such as appearance or athleticism, will help overweight or obese youngsters experience improved self-worth. Indeed, the personal experiences of overweight/obese youngsters are revealing in this context (Chapter 5). The severely obese young people (aged 4 to 18 years) interviewed by Foster and Page in Chapter 5 were keen to increase their competence at physical activity and sport, in part to overcome the physical difficulties that they experienced being obese and unfit. However, they also felt that the physical difficulties they faced when trying to exercise while obese tended to reinforce their obesity by limiting the activities they were capable of doing and thereby facilitating additional weight gain. A similar process of reinforcement was evident in the unrealistic targets for weight loss, which provoked "low self-motivation and depleted enthusiasm."

The role of parents in the emergence of childhood obesity is pervasive; obese parents tend to have obese offspring through either genotypic or phenotypic susceptibility. However, the majority of parents with obese children do not appear to understand either the causes of their child's obesity or what they can do to alleviate the problem (Chapter 6). Many parents seek professional help but are equivocal about the support they receive from health care professionals. The process of reassurance by health care professionals was also addressed in instances where parental concerns were downplayed by suggesting there wasn't a problem or that "these things all correct themselves" — an approach parents felt denied rather than acknowledged their concerns. Worse still, although many parents were initially made to feel "paranoid," some were ultimately made to feel inadequate or even responsible once the problem had finally been acknowledged. Parents offered treatment for their children found the treatment to be very basic, ineffective, and predominantly "negative" (i.e. "don't ..."), and some felt they were very much on their own when it came to looking for effective therapy. Edmunds suggests that the tendency for health care professionals to blame parents or children for obesity overlooks the broader contextual, structural, and cultural forces that are fueling the current incidence of childhood obesity. Whatever the causes, parents and children need the support of health care professionals and better understanding from society at large. Further research is needed to explore the attitudes and views of health care practitioners dealing with childhood obesity, to identify barriers to the provision of appropriate support, advice and care.

THE ROLE OF BIOLOGICAL AND SOCIAL PROCESSES IN THE ETIOLOGY OF CHILDHOOD OBESITY

The etiology of many diseases of lifestyle can best be investigated through birth cohort studies that have the statistical ability to unravel the complex interactions between candidate causative and associated variables (Chapter 7). The study of childhood obesity is particularly important in this context because of the known association of emergent obesity with critical periods during child growth (Cameron and Demerath 2002). The U.K. is host to the Avon Longitudinal Study of Parents and Children (ALSPAC), one of the world's most successful birth cohort studies; however, the challenges of long-term funding, participant attrition, and temporal changes in exposures and outcomes remain issues. Nonetheless, the opportunities for comparative analyses across cohorts have an important role to play in providing evidence on which action to prevent obesity can be based.

Despite the consensus that high-energy intake and low-energy expenditure are the causes of obesity, empirical evidence of precisely what intakes and expenditures are harmful is very difficult to accumulate. The main reason for this is the lack of precise measurement tools to assess dietary intake or energy expenditure (Chapter 8). An interesting consequence of the weaknesses of intake data is that preferences for food and activity may be more predictive of future outcomes than the intakes and actual physical activities undertaken. Clearly in this context, it is important to assess behaviors that are measurable with limited scope for imprecision and self-report bias. There is, for instance, little good evidence on the actual role of eating patterns, such as snacking, eating out of home, fast food consumption, and portion size, because of the influence of confounding factors and the problems with the assessment of food intake. The links to childhood obesity thus remain speculative. "Grazing" (the frequent consumption of small quantities of food) may be protective depending on what is being grazed and the timing of the introduction of eating patterns. The information on the age at which children begin to respond to portion size seems particularly relevant. The failure of previous dietary and activity prescriptions to effectively prevent childhood obesity bear testimony to the need for an appreciation of the multiple and interrelated determinants of children's behaviors.

Like food intake, there is no speculation about the importance of physical activity in the etiology of obesity. However, like food intake, its assessment is plagued with difficulties (Chapter 9). Indirect measurements using activity diaries flounder when the children fill them in sporadically and inaccurately. The choice of a specific day or days that truly represent "habitual" physical activity is also problematical, but the recent use of accelerometers has to some extent reduced some of these potential sources of inaccuracy. Such instrumentation may also be critical in identifying individual variation in the actual quantity and quality of physical activity within any particular activity period. A much greater understanding of human individual variability is needed before appropriate health messages can be effective (Chapter 10). Such variability in response to an exercise program may have its origins in behavioral factors such as noncompliance or in physiological factors such as compensating by increasing energy intake or decreasing nonexercise activity. Physically active individuals appear better able to regulate food intake and energy balance, but the evidence is that only a loose coupling exists between activity-induced energy expenditure and energy intake. Such evidence has potentially optimistic implications for the use of exercise in weight control as it means people do not automatically eat more as a result of exercise.

The notion that perinatal factors may affect the subsequent risk of obesity is relatively new, but the evidence is compelling (Chapter 11). The ALSPAC birth cohort study once again provides useful data to demonstrate the relationship between birth weight, early growth, and the development of risk factors by mid- to late childhood. A child born of low or lower birth weight who then grows rapidly during infancy is more likely to have increased weight, higher total body fat, and greater centralization of fat. These are outcomes that highlight increasing risk factors for obesity and adult morbidity encapsulated within the acquisition of an array of problems now called "Syndrome X." The tracking of BMI is crucial within the risk factors contributing toward adult morbidity. The 1958 British birth cohort study has been used to elucidate the trajectory of BMI between 7 and 42 years of age (Chapter 12). The development of multilevel modeling as a powerful statistical technique that could be applied to such data allowed the investigation of different factors affecting early life (social class, parental BMI, birth weight, infant feeding method) and childhood (childhood BMI at 7 and 11 years and pubertal maturation). The outcomes were BMI trajectories from 7 to 42 years (six time points) and from 16 to 42 years (four time points). It appears that the rate of change of BMI is significantly related to socioeconomic status (SES) in early childhood and then to parental BMI from mid-childhood onward. The relationship to SES was not explained by mother's BMI or smoking in pregnancy, birth weight, or breast-feeding and reflects the unexplained association between lower social group at birth and adult obesity in industrialized countries, which warrants further investigation.

THE PREVENTION AND TREATMENT OF CHILDHOOD OBESITY

The prevention of childhood obesity cannot be left to individuals or indeed families. It requires the community to take responsibility for the health and well being of its constituents. These communities range from the clinical community treating severe obesity within families to the school community aiming to improve the health of its pupils during their formative years (Chapter 13). Both scenarios have been tested through innovative initiatives in Leeds, Yorkshire. The "Watch It" clinics are attended regularly by whole families who receive basic advice about diet and physical activity and get the support and encouragement to tackle their obesity problems. The "APPLES" program was initiated 10 years ago as a health promotion exercise with the aim of developing a teaching package that would improve pupils' diets and increase their physical activity. In fact each school was encouraged to develop their own program dependent on local conditions. Meanwhile, the problems of monitoring the trends of increasing childhood obesity remain challenging. The "TREND" project involves a series of 13 "marker" schools in Leeds that act as monitoring centers for the whole community and thus act in a wider setting.

Prevention is, of course, better than cure, but critical appraisal of systematic reviews identifies a remarkable lack of good-quality evidence on high-quality interventions for the prevention of childhood obesity (Chapter 14). The reviews highlight weaknesses in the design, conduct, and reporting of pediatric obesity prevention randomized control trials (RCT). Most interventions in longer term RCTs have not been successful in reducing obesity risk. Existing approaches to prevention may not have a realistic prospect of success. Dietary behaviors may be less modifiable than sedentary behaviors, and most dietary variables are difficult to measure accurately. Most reviews conclude that inaccurate measures of diet and activity have obscured the identification of effective interventions.

The question of when to intervene has largely been neglected. There is, however, accumulating evidence that early life may be more important in the development of obesity in later life, and the evidence based on childhood obesity prevention has increased markedly in recent years. However, it remains extremely limited, and no successful interventions have been identified at present. In the absence of evidence, a cautious approach is recommended, which involves doing no harm, bringing about improvements in child health and development independent of obesity, and targeting modifiable behaviors that are likely to be related to obesity. To this can be added targeting behaviors that can be measured accurately and without self-report bias. From this perspective, reduced sedentary behaviors and sugar-sweetened drink consumption and increased breastfeeding should be targeted.

Changing childhood diets is an important part of the strategy to deal with obesity. Thus, the results of two systematic reviews that explore the factors responsible for the (in)effectiveness of interventions to promote fruit and vegetable intake among children are of relevance (Chapter 15). The first of these focuses on experimental studies (controlled trials) that sought to improve fruit and/or vegetable intake, knowledge, attitudes, and/or intentions among children aged 4 to 10 years. The second examined qualitative research exploring children's views about fruit and vegetables. Most of the trials used different school-based interventions that combined learning about the health benefits of fruits and vegetables with food preparation, tasting, "modeling" (observing others eating fruit and vegetables), or modifying the school environment (such as improving the availability of fruits and vegetables). Pooling the results of these trials suggested that the interventions were capable of increasing fruit and vegetable intake by, on average, close to one fifth of one portion per day. These studies found that children identified three aspects of food that were important to them: taste, choice, and excitement. Children distinguished between fruit, vegetables, and others and preferred foods on the basis of taste and reported that they did not like the taste of many of the foods promoted as "healthy" (particularly vegetables). Indeed, although the children liked the taste of fruit, they tended to view "healthy" as synonymous with poor taste. Thus, promoting fruit alongside vegetables as "healthy food" tended to undermine children's views of fruit as sweet and tasty. Children did not feel it was their role to be interested in health and felt that the health benefits of fruits and vegetables were exaggerated or irrelevant. They enjoyed being free to exercise choice and viewed eating foods their parents discouraged (such as sweets) as a way of asserting their independence. Likewise, children saw meal times as important opportunities for social interaction and found eating confectionary with friends "exciting." Although none of the experimental trials whose interventions matched children's views was harmful, the more successful interventions sought to improve both knowledge *and* access to fruits and vegetables, and included one or more of the three key aspects of children's views: treating fruit and vegetables separately, reducing the emphasis on health, and providing knowledge about, alongside access to, fruits and vegetables.

In view of the most recent U.K. government plans to restrict the advertising of food on television, it is germane to report on a systematic review of research exploring the nature of "food promotion" to children and its impact on children's knowledge, preferences, and behavior (Chapter 16). The studies suggest that food is promoted to children more than any other product (except for toys, and then only at Christmas time) and that food promotion focuses more on "fun" and "taste" than on "health" or "nutrition." Indeed, compared with recommended diets, the foods most actively promoted are higher in salt, sugar, and fat. Television is the principal advertising medium, although a wide range of "below-the-line" promotional techniques were also used (including novel packaging, tie-ins with films, and free gifts or tokens). Food advertising to children is dominated by five specific items: presugared breakfast cereals, confectionary, savory snacks, soft drinks, and fast-food outlets.

To assess whether food promotion techniques had any impact on children's food-related knowledge, preferences, or behavior, 33 experimental and observational (cross-sectional) studies were examined for evidence of potential or actual causality (according to the Bradford-Hill criteria). Although some of these studies were well-designed trials with sufficient sample sizes to provide robust statistical power, many were methodologically weak, and most could not disentangle food promotion per se from other potential confounders. These limitations aside, the studies reviewed do suggest that exposure to food promotion for "low nutrition foods" was associated with poorer nutritional knowledge (and vice versa). Likewise, there was evidence that food promotion was associated with significant changes in food preferences — with children preferring the food (or brand of food) that had been promoted. Finally, there is evidence that food promotion increased purchasing of promoted foods, independent of price, and increased consumption of promoted foods. Although several studies have found a relationship between hours spent watching television and health-related variables (such as BMI or blood cholesterol levels), only one could demonstrate a relationship with exposure to food advertising per se. In this study, it was acknowledged that it is difficult to unpick the wide range of complex and dynamic factors that influence food knowledge, preferences, and behavior while isolating the possible influence of just one variable — in this case promotion. Nonetheless, there is a link between advertising and children's food knowledge, preferences, behavior, and diet/health status, and these effects are independent of other influences, such as parental behavior and price. Most of the current campaigns promoting food to children are encouraging them to make less healthy choices and will have a deleterious impact on their dietary health.

The Loughborough symposium highlighted the evidence for an increasing prevalence of obesity in the U.K., its etiology in biological and social processes, and the problems faced by efforts at prevention and treatment. Notice was given of the catastrophe in public health, and the concomitant increased government spending, that will be the inevitable consequences of this eventual epidemic. Childhood obesity is, of course, just one of a number of child health issues facing populations in both developed and developing countries. Many have similar underlying factors that relate to economic transition (diet, activity, deprivation, etc.). It is clear that in order to develop strategic plans to contain these potential epidemics multifactorial and multi-agency approaches are required.

REFERENCES

National Audit Office, *Tackling Obesity in England,* The Stationery Office, London, 2001
  Health Survey for England, The Stationery Office, London, 2001.

Chief Medical Officer's Annual Report 2002, Department of Health, London, 2003.

Marshall, S., Biddle, S.J.H., Gorely, P.J., Cameron, N., and Murdey, I.D., Relationships
  between media use, body fatness and physical activity in children and youth:
  a meta analysis, *Int. J. Obes.,* 28, 1238–1246, 2004.

Cameron, N. and Demerath, E.W., Critical periods in human growth: relationships to
  chronic disease, *Yearbook Phys. Anthropol.* 45 159184, 2002.

PART I

ASSESSING THE SCOPE AND IMPACT OF CHILDHOOD OBESITY

1

CHILDHOOD OBESITY: ASSESSMENT AND PREVALENCE

Tim J. Cole

CONTENTS

INTRODUCTION

The inexorable rise in the prevalence of obesity in adults over the past few decades, followed more recently by a similar rise in children, has been alarming enough in its own right. Accruing evidence has linked obesity to morbidity and mortality from a wide variety of chronic diseases, including heart disease, hypertension, Type 2 diabetes, and most recently cancer, making obesity one of the most serious public health issues of our time.

This chapter aims to cover three aspects of the epidemiology of childhood obesity: definition and assessment, changes in prevalence over time, and the prediction of future obesity. The results imply two parallel approaches to body-mass index (BMI) assessment: epidemiological classification based on the International Obesity Task Force (IOTF) BMI cutoffs (Cole et al., 2000) and longitudinal assessment based on BMI centile crossing.

DEFINITION AND ASSESSMENT

When a condition such as obesity is increasing at the rate seen over the past few decades, it is obviously important to keep track of the "epidemic" by monitoring its prevalence over time in a range of different situations. There are other good reasons to monitor prevalence. First, by relating differences in obesity prevalence from one community to another with differences in the prevalence of lifestyle risk factors, one can tease out certain aspects of lifestyle that predispose to, or alternatively protect against, obesity. Second, by using prevalence data to test the impact of intervention programs, one can perhaps identify an effective program that leads to a reduction in prevalence.

The first requirement for prevalence studies is a clear and unambiguous definition of the condition. Obesity is at its simplest excess adiposity, thus a definition requires a way to measure adiposity and a cutoff to identify at what point adiposity becomes excessive.

Surprisingly this is almost impossible to achieve. For a start, adiposity is difficult to measure — current expensive laboratory-based methods like dual-energy x-ray absorptiometry and isotope dilution do not measure body fat in exactly the same way and thus give slightly different answers; in addition, they are impractical for large-scale population use. Bioelectrical impedance has its advocates, but offers little more information than anthropometry. Weight and height are the simplest and most widely available anthropometry, usually summarized as BMI (BMI = weight/height^2). However, this inexpensive and cheerful proxy for adiposity cannot distinguish between fat mass and lean mass, thus limiting its epidemiological value.

For large-scale studies in adults, the consensus is to measure adiposity using BMI, and the Garrow–Webster cutoffs of 25 and 30 kg/m^2 are widely agreed as lower limits for overweight and obesity, respectively (Garrow and Webster, 1985). Ideally, cutoffs that provide a sensitive and specific diagnostic test for the risk of later morbidity or mortality should be chosen. However, this is generally not possible as the risk rises essentially linearly with BMI. Therefore, shifting the cutoff slightly up increases the sensitivity of the test at the expense of the specificity, and the sum of sensitivity and specificity is broadly independent of the exact cutoff chosen. Garrow-Webster cutoffs were therefore chosen empirically to be convenient round numbers as much as objective proxies for disease risk.

For children, excess adiposity is superimposed on normal growth-related adiposity changes, adding another layer of complexity. The definition of obesity first needs to establish what is normal adiposity for age, and then the cut-off can be set relative to normal to identify obesity with high sensitivity and specificity. Ideally, the gold standard should again be long-term outcome, allowing children at risk of a later adverse event to be identified in advance. This risk depends critically on the age of the child;

therefore, knowledge of the relationship between adiposity and later out-come at different ages in childhood is necessary. In addition, the individual pattern of change in adiposity over time is important—a child who is fat early in life may have a risk in adult life different from a child who is thin in childhood but who becomes fat as a young adult. The current knowledge base linking child obesity to later outcome is completely inadequate to provide cutoffs based on this level of evidence.

The absence of an objective basis for setting the cutoffs coupled with the need to conveniently adjust adiposity for age in childhood resulted in an alternative approach of using age-adjusted BMI centiles. Conventionally, a growth centile chart is used to define abnormal growth, with an extreme centile on the chart acting as the cutoff to identify abnormality, By extension of this principle, BMI centiles such as the 85th, 91st, 95th. and 98th, which are based on U.S. or U.K. references, have been proposed to define overweight and/or obesity. However, the justification for any particular growth reference or centile is entirely ad hoc.

Altogether, the result has been entirely predictable—researchers have come up with their own personal definitions of child obesity, which are incompatible with other definitions. They choose a measure of weight-for-height, usually but not always BMI, a familiar growth reference, e.g., their own national reference, and a centile cutoff that seems suitable. Table 1.1 gives some examples of child obesity definitions from papers published over the past 7 years, and it is clear that they differ widely from each other. The problem with such definitions is that they all provide a percent prevalence, yet there is no way for novice users to realize that they cannot be compared with others.

Table 1.1 Various Definitions of Child Obesity Used in Recent Publications

Weight-Height Index	Criterion	Obesity Definition	Reference
Body weight	% ideal weight	> 120%	Chu, 2001
(Weight-9)/height$^{3.7}$	Value	Mean	Hughes et al., 1997
Weight for height	WHO z-score	> +2 z-scores	De Onis and Blossner, 2000
Weight for height	Thai centile	> 97th centile	Sakamoto et al., 2001
Body-mass index	U.S. CDC centile	> 95th centile	Troiano and Flegal, 1998
Body-mass index	U.K. z-score	Mean	Bundred et al., 2001

This highlights the need for an international definition of child obesity, authoritative enough for researchers to use worldwide, so as to provide prevalence rates that are directly comparable.

The IOTF recognized this problem of definition and called a workshop in 1997 to address it. Both the proceedings of the workshop (Bellizzi and Dietz, 1999) and more recently the IOTF international BMI cutoffs that were its main outcome (Cole et al., 2000) have been published. The IOTF cutoffs are based on the same principles as all the definitions in Table 1.1. IOTF uses BMI as the measure of adiposity but adjusts for age using centiles from an amalgam of six large national reference datasets (U.K., U.S., Singapore, Netherlands, Hong Kong, and Brazil) and uses "centile-like" cutoffs for overweight and obesity that are linked to the adult BMI cutoffs of 25 and 30 kg/m² at age 18 years. The final part is the main innovation; the cutoffs use BMI centiles that are chosen to pass through the adult cutoffs at age 18. This means that the centile is broadly unaffected by whether the reference children are fat or thin — the cut-off is 25 (or 30) at age 18 whatever their level of fatness.

Figure 1.1 illustrates the principle for the British reference data. The usual seven or nine centiles on the BMI chart are augmented with another

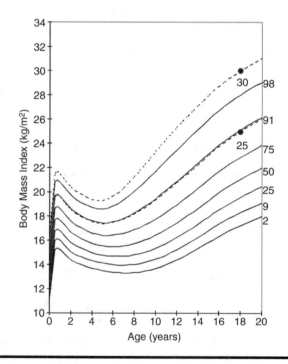

Figure 1.1   Body Mass Index Centiles for British Boys. The Centile Curves are Spaced Two Thirds of an SD Score Apart. Also Shown are Body Mass Index Values of 25 and 30 kg/m² at Age 18, With Extra Centile Curves Drawn Through Them.

two centiles passing through 25 and 30 kg/m² at age 18 years. This same process was followed for the other five datasets, leading to sets of six centile curves (one for each dataset) for overweight and obesity by sex. These four centile sets were then averaged by age to give pooled centile curves for overweight and obesity in boys and girls. The original reference (Cole et al., 2000) provides tables and plots of the separate centile curves and the averaged IOTF cutoffs.

CHANGES IN PREVALENCE OVER TIME

Since its publication in May 2000, the IOTF cutoffs' article has been cited over 300 times, reflecting both the interest in childhood obesity and the potential value of an international definition. To give a flavor of the scale of published work based on the cutoffs, studies are summarized here that have compared prevalence rates of overweight in the same population at two or three time points, allowing rates of increase in childhood obesity to be compared across countries.

Figure 1.2 gives details for overweight in boys from 10 countries (Chinn and Rona, 2001; Kalies et al., 2002; Kautiainen et al., 2002; Magarey et al., 2001; Tremblay et al., 2002; Wang et al., 2002). Corresponding data for obesity and for girls, omitted here for reasons of space, are available in the original articles. They tell broadly the same story. The highest prevalences and steepest rises are seen in the U.S. and Canada, whereas in China the prevalence is low and rising slowly and in Russia it is actually falling. This is assumed due to Russia's economic difficulties after the breakup of the

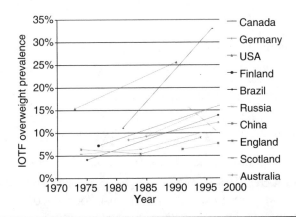

Figure 1.2   Trends Over Time in the Prevalence of Overweight in Boys From 10 Countries, Based on the IOTF International Cutoffs (Cole et al., 2000).

USSR in the early 1990s. The other countries are all rising at broadly the same rate, about 4% per 10 years. Although the most recent data in the figure are from 2000, indications elsewhere show that the rate has been increasing more rapidly since then. Either way, the figure indicates the wide variation in overweight prevalence in the 1980s and now and emphasizes the need for up-to-date information if intercountry comparisons are to be realistic.

PREDICTION OF FUTURE OBESITY

Which factors predict a child's risk of becoming obese later? A recent systematic review (Parsons et al., 1999) identified seven risk factors: parental fatness, social factors, birth weight, timing or rate of maturation, physical activity, dietary factors, and other behavioral or psychological factors. These can be grouped into three broad categories: genetic (parental fatness), physiological (birth weight and maturation), and behavioral (social, psychological, and lifestyle factors, including activity and diet). The genetic and physiological categories actually reflect an interaction between genes and the environment, although the way in which they interact is currently not clear. Programming during the fetal and early neonatal periods is likely involved, and this is now the focus of considerable work regarding the life course. Life course is the study of the way that patterns of growth and development through childhood relate to the risk of adverse adult outcome, in which obesity is known to play an important role.

Birth weight is widely documented as being inversely associated with later adverse outcomes such as heart disease, hypertension, and Type 2 diabetes, all conditions in which adult obesity is a risk factor. But a common pattern among low birth weight babies is rapid catch-up in growth soon after birth, and recent work suggests that it is the catch-up rather than the low birth weight that is important. Three papers so far have made the link, showing that weight gain in the period immediately after birth for 2 years (Ong et al., 2000), 4 months (Stettler et al., 2003), and 2 *weeks* (Singhal et al., 2003) is strongly associated with later obesity (Ong et al., 2000, Stettler et al., 2003) or insulin resistance (Singhal et al., 2003).

This relationship, if confirmed by other studies, has disturbing public health implications. The current view of parents, midwives, health visitors, and doctors is that babies growing fast in infancy are growing well—fast growth is seen as better than slow growth. But this latest work suggests that such a view is wrong and that infants need monitoring for too rapid growth just as much as for failure to thrive because long-term risks associated with excess growth are greater than previously perceived.

The other growth factor for later obesity identified by the systematic review was the child's rate of maturation. Girls who have an early menarche are more likely to be fat as adults, and this is seen as the impact of more rapid growth in the years leading up to menarche, as the child has to mature earlier. In addition many such girls are relatively fat in the period leading up to menarche, (Power et al., 1997), although the direction of causation is unclear. Are the girls fatter because they are on course for an early menarche, or has their greater fatness accelerated menarche? Both explanations are likely to have some truth.

Another important risk factor for later obesity, which the systematic review's terms of reference excluded, was the age of adiposity rebound, the second rise in BMI between 3 and 7 years of age. This has been known since 1984 to be a risk factor for later obesity, with an early age of adiposity rebound associated with a higher BMI later in childhood and adulthood (Rolland-Cachera et al., 1984). Some researchers have suggested that adiposity rebound is a "critical" period, in the sense that it is a time when growth programs the child's later obesity (Dietz, 1994). A recent study (Cole, 2004) has shown that the timing of the adiposity rebound can be interpreted in terms of BMI centile and BMI centile change. An early age of rebound corresponds to a high centile and/or upward centile crossing, and both of these growth patterns are statistically associated with a subsequent raised BMI. This is a consequence of the horse-racing effect (Peto, 1981), whereby children who are fat or becoming fat are more likely to be fat later. Therefore, the association of rebound with later fatness is statistical not physiological, ruling out the idea that the adiposity rebound is a critical period.

There is yet another physiological risk factor for later obesity, which is the child's height. The Bogalusa Heart Study has shown (Freedman et al., 2002) that taller children tend to be fatter as adults and that this association remains after adjustment for child BMI and triceps skinfold measurement.

Together, there are five distinct risk factors for later obesity that relate to child growth, as summarized in Table 1.2. Note that the two factors relating to timing have inverse associations, indicating positive associations with the

Table 1.2 Risk Factors for Later Obesity Relating to Growth During Childhood

Risk Factor	Association With Later Obesity	Interpretation
Birth weight	Positive	Upward weight centile crossing in pregnancy
Infancy weight gain	Positive	Upward weight centile crossing
Child height	Positive	Upward height centile crossing
Age of adiposity rebound	Negative	Upward BMI centile crossing
Timing of maturation	Negative	Upward BMI centile crossing

associated growth rate. All five factors indicate that rapid growth in childhood, whether in height or weight or BMI, is associated with a greater risk of obesity later. Birth weight is somewhat contradictory in that both high and low birth weight are risk factors. The Parsons systematic review highlighted the positive association, which may be mediated through gestational diabetes and macrosomia, but there is also the negative association relating to low birth weight and subsequent catch-up. This may mean that there is a U-shaped risk curve, with both extremes of birth weight at raised risk. Either way, rapid growth either in pregnancy (high birth weight) or neonatally (catch-up growth) is positively associated with later fatness.

There are two messages to take away from Table 1.2. The first is that *obesity is a growing problem*. This is a trite phrase describing the increasing prevalence of childhood obesity seen in recent years, but it is also a fundamental observation about the physiology underlying obesity in later childhood and adulthood. Children who are going to become obese later are likely to grow relatively fast at some point in childhood. This is plainly true for BMI, in that children have to cross centiles upward to be fat later. Surprisingly, however, it is also true for weight and height. The process of growing fast is associated with an imbalance in weight gain relative to height gain, with the result that such children also tend to get fat.

The second message from Table 1.2 is that the monitoring of growth, in particular BMI, is useful for spotting children most at risk of later fatness. This is already encouraged in the context of the adiposity rebound, but a more sensitive approach is to focus on longitudinal changes in BMI centile. As explained above, the information provided by the adiposity rebound is contained in the child's pattern of BMI centile change, but the centile information is predictive at all ages through childhood, not just at rebound.

In conclusion, patterns of childhood obesity over time need to be assessed in both the individual and the population, but different approaches are needed in the two cases. For epidemiological classification, prevalence rates of overweight and obesity need to be based on the IOTF BMI cutoffs (Cole et al., 2000). For longitudinal assessment in the individual, monitoring should be based on BMI centile crossing.

REFERENCES

Bellizzi, M. C. and Dietz, W. H., Workshop on childhood obesity: summary of the discussion, *Am. J. Clin. Nutr.*, 70, 173S–175S, 1999.

Bundred, P., Kitchiner, D., and Buchan, I., Prevalence of overweight and obese children between 1989 and 1998: population based series of cross sectional studies, *BMJ*, 322, 326–328, 2001.

Chinn, S. and Rona, R. J., Prevalence and trends in overweight and obesity in three cross sectional studies of British children, 1974–94, *BMJ,* 322, 24–26, 2001.

Chu, N. F., Prevalence and trends of obesity among school children in Taiwan—the Taipei children heart study, *Int. J. Obes.,* 25, 170–176, 2001.

Cole, T. J., Children grow and horses race: is the adiposity rebound a critical period for later obesity?, *BMC Pediatr.,* 4, 6, 2004.

Cole, T. J., Bellizzi, M. C., Flegal, K. M., and Dietz, W. H., Establishing a standard definition for child overweight and obesity: international survey, *BMJ,* 320, 1240–1243, 2000.

De Onis, M., and Blossner, M., Prevalence and trends of overweight among preschool children in developing countries, *Am. J. Clin. Nutr.,* 72, 1032–1039, 2000.

Dietz, W. H., Critical periods in childhood for the development of obesity, *Am. J. Clin. Nutr.,* 59, 955–959, 1994.

Freedman, D. S., Khan, L. K., Mei, Z. G., Dietz, W. H., Srinivasan, S. R., and Berenson, G. S., Relation of childhood height to obesity among adults: The Bogalusa Heart Study, *Pediatrics,* 109, U22–U28, 2002.

Garrow, J. S. and Webster, J., Quetelet's index (w/h^2) as a measure of fatness, *Int. J. Obes.,* 9, 147–153, 1985.

Hughes, J. M., Li, L., Chinn, S., and Rona, R. J., Trends in growth in England and Scotland, 1972 to 1994, *Arch. Dis. Child.,* 76, 182–189, 1997.

Kalies, H., Lenz, J., and Von Kries, R., Prevalence of overweight and obesity and trends in body mass index in German pre-school children, 1982–1997, *Int. J. Obes. Relat. Metab. Disord.,* 26, 1211–1217, 2002.

Kautiainen, S., Rimpela, A., Vikat, A., and Virtanen, S. M., Secular trends in overweight and obesity among Finnish adolescents in 1977–1999, *Int. J. Obes.,* 26, 544–552, 2002.

Magarey, A. M., Daniels, L. A., and Boulton, T. J. C., Prevalence of overweight and obesity in Australian children and adolescents: reassessment of 1985 and 1995 data against new standard international definitions, *Med. J. Aust.,* 174, 561–564, 2001.

Ong, K. K., Ahmed, M. L., Emmett, P. M., Preece, M. A., and Dunger, D. B., Association between postnatal catch-up growth and obesity in childhood: prospective cohort study, *BMJ,* 320, 967–971, 2000.

Parsons, T. J., Power, C., Logan, S., and Summerbell, C. D., Childhood predictors of adult obesity: a systematic review, *Int. J. Obes.,* 23 *Suppl* 8, S1–107, 1999.

Peto, R., The horse-racing effect, *Lancet,* 2, 467–468, 1981.

Power, C., Lake, J. K., and Cole, T. J., Body mass index and height from childhood to adulthood in the 1958 British born cohort, *Am. J. Clin. Nutr.,* 66, 1094–101, 1997.

Rolland-Cachera, M. F., Deheeger, M., Bellisle, F., Sempé, M., Guilloud-Bataille, M., and Patois, E., Adiposity rebound in children: a simple indicator for predicting obesity, *Am. J. Clin. Nutr.,* 39, 129–135, 1984.

Sakamoto, N., Wansorn, S., Tontisirin, K., and Marui, E., A social epidemiologic study of obesity among preschool children in Thailand, *Int. J. Obes.,* 25, 389–394, 2001.

Singhal, A., Fewtrell, M., Cole, T. J., and Lucas, A., Low nutrient intake and early growth for later insulin resistance in adolescents born preterm, *Lancet,* 361, 1089–1097, 2003.

Stettler, N., Kumanyika, S. K., Katz, S. H., Zemel, B. S., and Stallings, V. A., Rapid weight gain during infancy and obesity in young adulthood in a cohort of African Americans, *Am. J. Clin. Nutr.,* 77, 1374–1378, 2003.

Tremblay, M. S., Katzmarzyk, P. T., and Willms, J. D., Temporal trends in overweight and obesity in Canada, 1981–1996, *Int. J. Obes.*, 26, 538–543, 2002.

Troiano, R. P. and Flegal, K. M., Overweight children and adolescents: description, epidemiology, and demographics, *Pediatrics*, 101, 497–504, 1998.

Wang, Y. F., Monteiro, C., and Popkin, B. M., Trends of obesity and underweight in older children and adolescents in the United States, Brazil, China, and Russia, *Am. J. Clin. Nutr.*, 75, 971–977, 2002.

2

GLOBAL PERSPECTIVES ON ADOLESCENT OBESITY

Penny Gordon-Larsen and Barry M. Popkin

CONTENTS

INTRODUCTION

Over the past 15 years, there is increasing evidence that the structure of dietary intakes and the prevalence of obesity around the developing world have been changing at an increasingly rapid pace (Popkin, 2002).* In many ways, these shifts are a continuation of large-scale changes that have occurred repeatedly over time; however, we will assert and show that the changes now occurring in low- and moderate-income countries appear to be very rapid. Although initially these shifts were felt to be limited to higher-income urban populations, it is increasingly clear that these are much broader trends affecting all segments of society.

In this chapter, we will briefly explore several themes. The first is the general shift of obesity representing a global problem rather than one centered in a few high-income countries. The second is the rapid increase in obesity found in lower- and middle-income developing countries—a rate of change that appears to be greater than that found in higher-income countries. The third is the shift in the burden of obesity toward the poor on a worldwide basis. We then present the limited comparable data on trends in adolescent obesity across the globe, with specific attention to the link between obesity and economic and social development.

Finally, we provide, via use of a nationally representative longitudinal survey of U.S. adolescents, insights into the patterns of obesity changes over the adolescent and young adult period in the U.S.

GLOBAL OBESITY

Shifts in Adult Obesity Are Occurring Across the Globe

In a series of papers, the current levels of overweight in countries as diverse as Mexico, Egypt, and South Africa are shown to be equal or greater than those in the U.S. (Popkin, 2002a). Moreover, the rate of change in obesity in lower- and middle-income countries is shown to be much greater than in higher-income countries (see Popkin, 2002a, for the overview). Figure 2.1 presents the levels of obesity and overweight in several illustrative countries (Brazil, Mexico, Egypt, Morocco, South Africa, Thailand and China). Most interesting is the fact that many of these countries have quite high overweight levels but are very low income. Moreover, it probably surprises many people that the levels of obesity of several countries — all with much lower income levels than the U.S. — are so high.

---

* This chapter focuses on obesity patterns and trends. Elsewhere, we have previously presented comparable data on dietary and physical activity change (Bell et al., 2002; Gordon-Larsen et al., 2002; Popkin, 2002b and 2003).

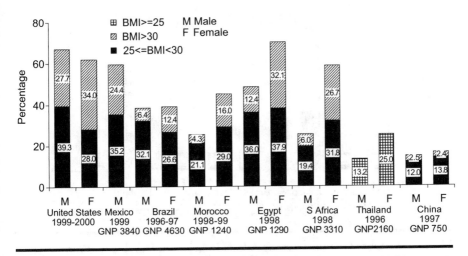

Figure 2.1 Obesity Patterns Across the Developing World. (From Popkin, B.M., *Public Health Nutr.*, 5, 93–103, 2002a).

There is enormous heterogeneity in the patterns, trends, and timing of obesity among developing countries. Many countries in Latin America began their transition earlier in the past century and certainly entered the nutrition-related noncommunicable disease (NR-NCD) stage of the transition far earlier than did other regions; however, other countries such as Haiti and subpopulations in Central America are still in the receding famine period. Moreover, some countries, such as Mexico, experienced an accelerated transition in the 1990s (Rivera et al., 2002). The Middle East, North Africa, and Asia appear to have begun their transition at a much later date as have most other countries in the developing world except for the Western Pacific nations.

Changes in the Developing World Are Faster Than in Higher-Income Countries

Figure 2.2 shows how quickly overweight and obesity status have emerged as major public health problems in some of these countries. Compared with the U.S. and European countries, where the annual increase in the prevalence of overweight and obesity among adult men and women is about 0.25 for each, the rates of change are very high in Asia, North Africa, and Latin America—two to five times greater than in the U.S.

The Burden of Obesity Has Shifted to the Poor

A large number of low- and moderate-income countries have a greater likelihood that adults residing in lower-income or lower-educated households

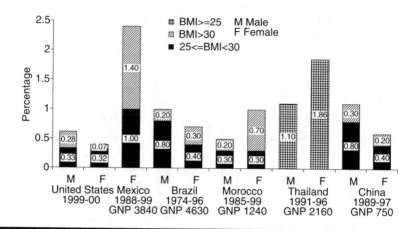

Figure 2.2   Obesity Trends Among Adults in Selected Developing World Countries (The Annual Percentage Point Increase in Prevalence). (From Popkin, B.M., *Public Health Nutr.*, 5, 93–103, 2002a).

are overweight and obese relative to adults in higher-income or higher-education households (Monteiro et al., in press). This study, based on multilevel analyses of 37 nationally representative data sets, shows that countries with a GNP per capita over about $1700 are likely to have a burden of obesity greater among the poor. We applied multilevel logistic regression analyses on the risk of obesity [body mass index (BMI) 30 kg/m²] to anthropometric and socioeconomic data collected by nationally representative cross-sectional surveys on women aged 20 to 49 ($n = 148,579$) conducted from 1992 to 2000 in 37 developing countries within a wide range of world regions and stages of economic development (GNP from $190 to $4440 per capita). Importantly, these analyses demonstrate the differential effect of an individual's socioeconomic status (SES) within a country's GNP level.

Belonging to the lower SES group confers strong protection against obesity in low-income economies, can reduce or increase obesity in lower-middle income economies, and is a systematic risk factor for obesity in upper-middle income developing economies. A multilevel logistic model, which included an interaction term between the country's GNP and each woman's SES, indicates that obesity starts to fuel health inequities in the developing world when the GNP reaches a value of about $2500 (U.S.) per capita (see Figure 2.3). Examples of countries above the $2500 per capita income level and defined as upper-middle income developing economies that have higher obesity among the lower SES groups include Mexico, Brazil, Turkey, and South Africa.

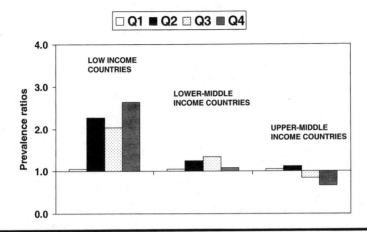

Figure 2.3 Prevalence ratios for women's obesity, for women aged 19–49 years, Demographic and Health Surveys, by quartiles of years schooling in 37 developing countries (1992–2002). Source: Monteiro et al. 2004(b).

Adolescent Obesity Trends

A study of adolescent obesity dynamics was undertaken across four longitudinal studies (Wang et al., 2002), using nationally representative data from Brazil (1975 and 1997), Russia (1992 and 1998), and the U.S. [NHANES I (1971–1974) and NHANES III (1988–1994)] and a nationwide survey data from China (1991–1997). To define overweight, we used the sex–age-specific BMI cutoffs recommended by the International Obesity Task Force (IOTF). The sex–age-specific BMI 5th percentile based on the U.S. NHANES I data was used to define underweight.

Overweight prevalence measured by changes in the time periods noted above (Figure 2.3) has increased in Brazil (4.1% to 13.9%), China (6.4% to 7.7%), and the US (15.4% to 25.6%). In Russia, overweight decreased (15.6% to 9.0%). The annual increase rates of overweight prevalence (percentage points) are: 0.5 percentage points (Brazil), 0.2 percentage points (China), −1.1 percentage points (Russia), and 0.6 percentage points (US). These are based on nationally representative weighted averages for Brazil, Russia and the US but do not adjust for any population composition changes.

The current prevalence of overweight varies substantially across the four countries that we examined and seems to show a large increase vs. the changes experienced by adults in these countries. The burden of nutritional problems is shifting from energy imbalance deficiency to excess among older children and adolescents in Brazil and China, which perhaps is related to changes in key environmental factors across countries. For example, the GDP tripled in the U.S. and Brazil and increased greater than 10-fold in China during this time (e.g., rise of living standards, including an increase of food production and consumption, along with declines in physical activity and increases in inactivity, such as TV ownership). Conversely, Russia saw a worsening of the economy and a decline in living standards (e.g., decline of living standards, including a decrease of food consumption and production). Over the past three decades, the percentage of older children and adolescents who were overweight tripled in Brazil and almost doubled in the U.S.

Similar to adults, the changes in youth obesity in these four countries vary across levels of household income. Over the past two decades, the prevalence of overweight in Brazil increased most quickly in high-SES groups; however, in the U.S., the increase was greatest in low-SES groups (see Figure 2.4). In addition, the prevalence of overweight was much higher in urban areas than in rural areas in Russia and the U.S.

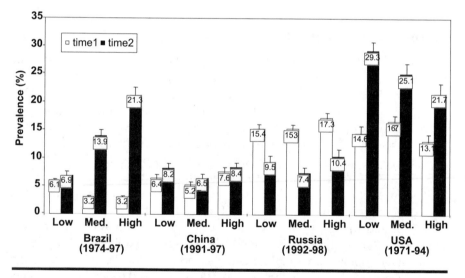

Figure 2.4   Relationship Between Household Income Per Capita and the Prevalence of Overweight (Defined Using IOTF Cutoff Points). Prevalence % are in Boxes, and T-bars Show Standard Errors. Income Tertiles were Used in Each Country to Indicate Low, Medium, and High Levels. For the First and Second Surveys, Respectively, *n* = 56,295 and 4875 for Brazil, 3014 and 2688 for China, 6883 and 2152 for Russia, and 4472 and 6108 for the U.S. (From Wang, Y., et al., *Am. J. Clin. Nutr.,* 75, 971–977, 2002).
Reproduced with permission by the American Journal of Chemical Nutrition.

Childhood and Adolescent Obesity Patterns Across Europe

The prevalence of obesity has also risen dramatically in many countries across Europe (Lobstein and Frelut, 2003; Guillaume and Lissau, 2002). Among children (ages 7 to 11 years), overweight, defined using the IOTF cutoff points, differs across Europe from France (19%) and the U.K. (20%), Sweden (18%), and Denmark (15%) in Western and Northern Europe to Russia (10%) and Poland (18%) in the Eastern bloc, and to Spain (34%), Italy (36%), and Malta (35%) in Southern Europe (Guillame and Lissau 2002). Adolescent (14 to 17 years) overweight patterns are similar, with particularly high levels in the U.K. (21%) and Southern Europe, including Spain (21%), Greece (22%), and Cyprus (23%). Guillame and Lissau (2002) postulate many reasons for these variations. For example, the North–South trend is clearly seen, and the lower prevalence in Central and Eastern Europe occurs in countries whose economies suffered from recessions during the economic and political transition of the 1990s.

Across all regions of the world, it seems that the levels of adult obesity far exceed those of children. However, it also seems that child and adolescent obesity have rapidly increased in recent years.

Shifts in Age-Specific Time Trends

There are few longitudinal studies that follow different age groups at the same time using a mixed cohort study. Studies of both dietary and body composition trends seem to indicate an increase in rates of change across the world. This increase has also been found in studies of the impact of income and price changes on the structure of the Chinese diets as well as studies on obesity trends in the U.S. In China, two studies have shown that increased consumption of higher fat foods and higher fat diets have accelerated in conjunction with changes in income (Du et al., 2004; Guo et al., 2000). In each case, longitudinal analysis was used to show a significant increase in the impact of income on diet.

A longitudinal study in which over 9000 U.S. young adults born between 1957 and 1964 were resurveyed 13 times over the span of two decades found that more than 25% were obese by age 35 (McTigue et al., 2002). Timing of obesity onset was highest among black females, moderate in Hispanic females, and lowest in white females, while among males highest onset was seen for Hispanic males. In addition, McTigue et al. (2002) found a large secular weight trend: people born in later calendar years tended to have larger age-specific BMI. In every case in which the same age was sampled for both birth cohorts, the 1957-born group had a mean BMI lower than the 1964-born group. The intermediate birth cohorts had intermediate-range mean BMI values.

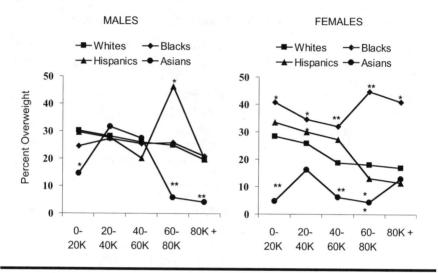

Figure 2.5 Overweight Prevalence by Category of Income in U.S. for White, Black, Hispanic, and Asian Adolescents, Using Nationally Representative Data From the National Longitudinal Study of Adolescent Health; *n* = 13,113. *Statistically Significant Differences From Same-Sex, Non-Hispanic Whites, *p* < .05. **Statistically Significant Differences From Same-Sex, Non-Hispanic Whites, *p* < .01. (From Gordon-Larsen, P., et al., *Obes. Res.*, 11, 121–129, 2003).

Relationship Between Income and Overweight Among U.S. Adolescents

The relationship between socioeconomic correlates of overweight and overweight (BMI 95th percentile, CDC 2000 growth curves) were examined using a nationally representative sample of over 20,000 U.S. adolescents (The National Longitudinal Study of Adolescent Health). In this study, Gordon-Larsen et al. (2003) found a clear difference in the shape of the income and overweight relationship, particularly for females. A clear inverse relationship between higher income and lower overweight was seen only among non-Hispanic white females (see Figure 2.5). In females, black—white disparity in overweight actually increased at the highest income levels.

U.S. CASE STUDY–INSIGHTS INTO CURRENT DYNAMICS IN THE HIGHER-INCOME WORLD

Gordon-Larsen et al. (2004) sought to examine patterns of change in obesity among U.S. white, black, Hispanic, and Asian teens as they transition from adolescence to young adulthood. At baseline (1996), obesity prevalence (using IOTF cutoff points) was 10.9%. Five years later, a substantial proportion

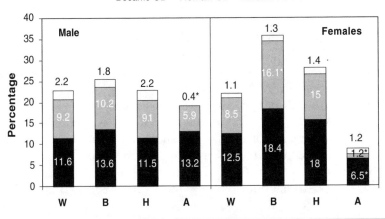

■ Became OB ▨ Remain OB ▢ Became Non-OB

Figure 2.6 Shifts in Obesity From Adolescence (1995) to Young Adulthood (2001) in White, Black, Hispanic, and Asian U.S. Adolescents and Adults Using Nationally Representative, Longitudinal Data From the National Longitudinal Study of Adolescent Health; $n = 9795$. *Statistically Significant Differences From Same-Sex, Non-Hispanic Whites, p # .003. (From Gordon-Larsen, P., et al., Am. J. Clin. Nutr., 80, 569–575, 2004).

of the adolescents who were nonobese at baseline had become obese (12.7% incidence) and remained obese (9.4% maintenance), with a small proportion moving from obese to nonobese (1.6% reversal). Obesity incidence was especially high among non-Hispanic black (18.4%) and Hispanic females (15.8%) relative to white females (12.5%) as was obesity maintenance among black (16.1%) and Hispanic (10.9%) females relative to white females (8.5%) (Fig. 2.6). Importantly, these data represent approximately 15.1 million 13- to 20-year-old U.S. school children. Thus, in the 5-year study period, more than 1.9 million adolescents became obese, more than 1.3 million adolescents remained obese, and only 241,653 adolescents became nonobese.

These results mirror many smaller studies that display a significant tendency for childhood and adolescent overweight to persist into adulthood (Serdula et al., 1993; Power et al., 1997; Srinivisan et al., 1996). The Gordon-Larsen et al. (2004) findings indicate that the transition between adolescence and young adulthood appears to be a period of increased risk for development of obesity. This upward trend is evident in both males and females and in all major U.S. ethnic groups, particularly in non-Hispanic blacks. The trend foreshadows higher rates of diabetes and nutrition-related chronic degenerative diseases emerging at younger ages (Mokdad et al., 2001; 2003) and underscores the major need for preventive strategies to curb this trajectory.

CONCLUSIONS

In this chapter, we have shown a profound global shift in obesity. There are clear associations among global economies, recession, and economic factors and obesity in adults, adolescents, and children. The high obesity prevalence in adolescents is shown to be persistent into adulthood, with high incidence shown in the transition to adulthood. The global obesity problem requires global obesity preventive and treatment solutions. Greater attention must be paid to research on economic, environmental, and social determinants of obesity in children and adolescents. We see from the U.S. data that the increase in obesity from adolescence to adulthood is tremendous. Other countries following this trajectory are likely to see substantial adult obesity and its associated co-morbidities if this trajectory is not curbed.

REFERENCES

Bell, A.C., Ge K., and Popkin, B.M., The road to obesity or the path to prevention? Motorized transportation and obesity in China, *Obes. Res.*, 10, 277–283, 2002.

2000 CDC Growth Charts: United States. Centers for Disease Control and Prevention, National Center for Health Statistics, *available at* http://www.cdc.gov/growthcharts (accessed 21 April 2003).

Du, S., Mroz, T.A., Zhai, F., and Popkin, B.M., Rapid income growth adversely affects diet quality in China—particularly for the poor! *Soc. Sci. Med.*, 59, 1505–1515, 2004.

Gordon-Larsen, P., Adair, L.S., and Popkin, B.M., Ethnic differences in physical activity and inactivity patterns and overweight status: The National Longitudinal Study of Adolescent Health, *Obes. Res.*, 10, 141–149, 2002.

Gordon-Larsen, P., Adair, L.S., and Popkin, B.M., The relationship between ethnicity, socioeconomic factors and overweight: The National Longitudinal Study of Adolescent Health, *Obes. Res.*, 11, 121–129, 2003.

Gordon-Larsen, P., Adair, L.S., Nelson, M.C., and Popkin, B.M., Five-year obesity incidence in the transition period between adolescence and adulthood: The National Longitudinal Study of Adolescent Health, *Am. J. Clin. Nutr.*, 80, 569–575, 2004.

Guillame, M. and Lissau, I., Epidemiology, in *Child and Adolescent Obesity: Causes and Consequences, Prevention and Management*, Burniat W., Cole, T., Lissau, I., and Poskitt, E.M.E., Eds., Cambridge University Press, Cambridge, pp. 28–49, 2002.

Guo, X., Mroz, T.A., Popkin, B.M., and Zhai, F., Structural changes in the impact of income on food consumption in China, 1989–93, *Econ. Dev. Cult. Change* 48, 737–760, 2000.

Lobstein, T. and Frelut, M.-L., Prevalence of overweight among children in Europe, *Obes. Rev.*, 4, 195–200, 2003.

McTigue, K.M., Garrett, J.M., and Popkin, B.M. The natural history of obesity: weight change in a large U.S. longitudinal survey, *Ann. Int. Med.*, 136, 857–864, 2002.

Mokdad, A.H., Bowman, B.A., Ford, E.S., Vinicor, F., Marks, J.S., and Koplan, J.P., The continuing epidemics of obesity and diabetes in the United States, *JAMA*, 286, 1195–1200, 2001.

Mokdad, A.H., Ford, E.S., Bowman, B.A., Dietz, W.H., Vinicor, F., Bales, V.S., and Marks, J.S., Prevalence of obesity, diabetes, and obesity-related health risk factors, 2001, *JAMA,* 289, 76–79, 2003.

Monteiro, C.A., Conde, W.L., Lu, B., and Popkin, B.M., Obesity inequities in health in the developing world? *Int. J. obes. Relat. Metab. disord.,* 28, 1181–1186, 2004.

Popkin, B.M., An overview on the nutrition transition and its health implications: the Bellagio Meeting, *Public Health Nutr.,* 5, 93–103, 2002a.

Popkin, B.M., The dynamics of the dietary transition in the developing world, in *The Nutrition Transition: Diet and Disease in the Developing World,* Caballero, B. and Popkin, B.M., Eds., Academic Press, London, pp. 111–128, 2002b.

Popkin, B., The nutrition transition in the developing world, *Dev. Policy Rev.,* 21, 581–597, 2003.

Power, C., Lake, J.K., and Cole, T.J., Measurement and long-term health risks of child and adolescent fatness, *Int. J. Obes. Relat. Metab. Disord.,* 21, 507–526, 1997.

Rivera, J.A., Barquera, S., Campirano, F., Campos, I., Safdie, M., and Tovar, V., Epidemiological and nutritional transition in Mexico: rapid increase of non-communicable chronic diseases and obesity, *Public Health Nutr.* 5, 113–122, 2002.

Serdula, M.K., Ivery, D., Coates, J.R., Freedman, D.S., Williamson, D.F., and Byers, T., Do obese children become obese adults? A review of the literature, *Prev. Med.* 22, 167–177, 1993.

Srinivisan, S.R., Bao, W., Wattigney, W.A., and Berenson, G.S., Adolescent overweight is associated with adult overweight and related multiple cardiovascular risk factors: The Bogalusa Heart Study, *Metab. Clin. Exp.* 45, 235–240, 1996.

Wang, Y., Monteiro, C., and Popkin, B.M., Trends of overweight and underweight in children and adolescents in the United States, Brazil, China, and Russia, *Am. J. Clin. Nutr.* 75, 971–977, 2002.

3

A BRIEF REVIEW OF THE HEALTH CONSEQUENCES OF CHILDHOOD OBESITY

Fiona Regan and Peter Betts

CONTENTS

INTRODUCTION

Obesity is an ever-increasing problem in today's society for both children and adults. The World Health Organization declared obesity as a global epidemic in 1998 (World Health Organization, 1998), recognizing its long-term implications on health. The problem continues to increase today in the U.K. In Leeds, the prevalence has risen in primary school children such that in 1999 one in five 9-year-old and one in three 11-year-old girls were overweight (Rudolf et al., 2001). Local data from Southampton has demonstrated an ongoing increase in overweight 5 year olds, from 11% in 1993 to 15% in 2001, and this currently stands at 18% for those aged 5–7 years in 2003.

Body mass index (BMI, in kg/m^2) is the most commonly used index to delineate the prevalence of overweight and obesity in children because of its ease of measurement, but this calculation is likely to underestimate central obesity (McCarthy et al., 2003), a major marker for insulin resistance; hence, even on the current prevalence figures, we may not fully appreciate the potential significance of this problem.

Among all of the many children with obesity, it is essential to identify those with pathological and treatable causes, such as hypothyroidism or Cushing's syndrome, and those with genetic disorders, which may present in children <2 years of age. In reality, however, these cases are rare and if a child is tall and of normal IQ, investigation will rarely identify any on-going causative pathology.

Tracking studies have shown that an overweight child is likely to become an overweight adult (Whitaker et al., 1997; Sinaiko et al., 1999; Must et al., 1999; Serdula et al., 1993; Power et al., 1997; Guo and Chumlea, 1999; Mossberg, 1989). This is especially so if one parent is obese or if the child is already obese and especially so in the older child entering adolescence.

Does this year on year increase in obesity matter and what are the implications for both current health and the future as adults? This chapter will show that the current and future health consequences of obesity are grave and varied.

CONSEQUENCES OF OBESITY DURING CHILDHOOD

Increasing Insulin Resistance

Obese children have increasing evidence of insulin resistance, resulting in hyperinsulinemia as a compensating mechanism to maintain euglycemia. When the degree of resistance exceeds the secretory capacity of the islet cells, then glucose intolerance and later Type 2 diabetes occur. Sinha et al. (2002) recently reported a >20% incidence of glucose intolerance in very obese children and adolescents with a 4% incidence of unrecognized diabetes.

Diabetes Mellitus

Type 2 diabetes mellitus is more strongly associated with obesity in adults than any other medical condition (Must et al., 1999). Historically, this was previously described as "adult onset diabetes." However, the incidence of both Type 1 and Type 2 diabetes in childhood has increased markedly over the past 10 years. In Cincinnati, there has been a 10-fold rise in Type 2 diabetes in young people aged 10–19 years, from 0.7/100,000 per year in 1982 to 7.2/100,000 in 1994, accounting for 33% of newly diagnosed diabetic children in Cincinnati in 1994 (Pinhas-Hamiel et al., 1996). The mean age at presentation was 13.8 years and mean BMI was 37.7 kg/m^2, with 38% having a BMI >40 kg/m^2. In the U.S., Type 2 diabetes is now reported to account for up to 45% of new cases of diabetes in childhood (Aye and Levitsky, 2003).

Type 2 diabetes has been reported in the U.K. to be associated with obesity but not in such large numbers (Drake et al., 2002). In Leeds in 2000, only 5% of the diabetic population under the age of 30 years was identified as having Type 2 diabetes (Feltbower et al., 2003).

The presence of diabetes carries risks of micro- and macrovascular complications, and this may be more so in Type 2 diabetes than in Type 1. Cardiovascular risks are particularly high, given that both obesity and hyperinsulinemia are known to be cardiovascular risk factors in their own right (Despres et al., 1996).

As Type 2 diabetes occurs with an increased frequency and starts to appear at an ever decreasing age, we face the possibility that, in addition to diabetes, coronary heart disease may become a disease of young adulthood.

Type 2 diabetes is reversible and potentially preventable if it is recognized before beta cell failure has occurred. It is imperative that we diagnose this condition early and endeavour to get young people to lose weight before permanent islet cell damage.

Recent evidence shows that weight also has a significant effect on the increase in Type 1 diabetes that has been seen in children. The age of presentation of Type 1 diabetes has been decreasing in the pediatric population at the same time as weight has increased. This led Wilkin (2001) to propose a causal relationship between the two, with weight accelerating its earlier onset, termed the "accelerator hypothesis." BMI SDS at diagnosis, as an index of obesity, and weight SDS change from birth were inversely related to age at presentation, with heavier children presenting at a younger age (Kibirige et al., 2003). In Southampton, we found that over the past 20 years in the pediatric population BMI SDS at diagnosis of Type 1 diabetes has increased with time and BMI SDS 6 months after diagnosis, when weight has been regained, is inversely related to age at diagnosis. In addition, those children with Type 1 diabetes had evidence of insulin resistance with increased abdominal circumference as a proxy for abdominal fat (Betts et al., 2005). The consequences of the earlier presentation of both Type 1 and Type 2 diabetes resulting in increased duration of diabetes, and hence increased risk and/or earlier emergence of complications, has wide-reaching implications for the health economy.

Cardiovascular Risks in Association With Obesity

The association between adult obesity and cardiovascular risk has been well established (Rexrode et al., 1996; Kannel et al., 1996; Hubert et al., 1983). There is now an emerging wealth of evidence demonstrating that childhood obesity is also associated with adverse cardiovascular risk factors. These risk factors include raised fasting lipids, hypertension, and increased left ventricular mass. As in the adult population, central fat mass is a more important independent correlate in children than percent body fat (Daniels et al., 1999). The risks may be of particular importance in certain ethnic groups. Whincup et al. (2002) found that British South Asian children have proportionally higher triglycerides, fibrinogen, and insulin levels compared with British white children of the same adiposity. This is important if targeting prevention at high-risk groups is to be advocated.

Increase in left ventricular mass is associated with childhood obesity (Daniels et al., 1995; Daniels et al., 1999; Gutin et al., 1998). Left ventricular function may also be seen as an association between lower midwall fractional shortening [a known cardiovascular risk factor (De Simone et al., 1996)] and percent body fat (Gutin et al., 1998). It is possible these changes are related to hyperinsulinemia, which has been implicated in the etiology of raised blood pressure (Gutin et al., 1998).

Hypertension in childhood is not common but is seen in 20 to 30% of children with a weight >120% of ideal body weight (Figueroa-Colon et al., 1997).

Change in weight centile during childhood appears to have particular relevance with regard to these risk factors. The Minneapolis Children's Blood Pressure Study delineated that increases in weight and BMI during childhood, but not height, are significantly correlated with young adult levels of insulin, lipids, and systolic blood pressure (Sinaiko et al., 1999). Thus, adverse risk factors are probably due to weight gain in excess of that needed for growth. Furthermore, if we can reduce the rate of weight gain in these children, we may be able to alter the course of potential adult morbidity.

The clustering of a range of cardiovascular adverse risk factors in obese children is perhaps of greatest concern because it leads to a multifaceted risk profile. Analyses of the data from the Bogalusa Heart Study have shown that obese children not only have a raised odds ratio for many adverse factors but also have marked clustering of these factors (Freedman et al., 1999). Children with a BMI >95th centile, compared with children with <85th centile, had an odds ratio of 2.4 for raised cholesterol, 2.4 for diastolic hypertension, 3.0 for raised low-density lipids (LDL), 4.5 for systolic hypertension, 7.1 for raised triglycerides, and 12.6 for raised fasting insulin. Overweight children were 9.7 times more likely to have two risk factors and 43.5 times more likely to have three risk factors.

PUBERTY AND RELATED DISORDERS

Overweight children tend to be taller and have an advanced bone age. Earlier puberty may occur in girls, resulting in an earlier age of menarche. In addition, early maturing girls are more also likely to be obese during adolescence, with this likelihood persisting into adulthood (Wang, 2002; Adair and Gordon-Larsen, 2001; Beunen et al., 1994; Kaplowitz et al., 2001).

Obese boys on the other hand have a considerable variation in timing of puberty. Many may have delayed puberty, as demonstrated by a recent study showing a negative association of male obesity and sexual maturation (Wang, 2002). Obese pubertal boys often have significant gynecomastia (Voors et al., 1981).

Polycystic ovary syndrome (PCOS) is a common condition that frequently has its origin in adolescence (Rosenfield, 1990). It is more common in obese female adolescents, and, although the pathophysiology is complex, it is related to hyperinsulinemia, insulin resistance, and hyperandrogenism. Some of the classical features of PCOS, such as hirsuitism, anovulatory cycles, and menstrual irregularities, are more common in obese than in lean girls, and correlations between free androgen index and weight/height (Wabitsch et al., 1995) and testosterone and BMI (Malecka-Tandera et al., 1998) have been shown.

RESPIRATORY DISEASE

Over the past three decades, the prevalence of childhood asthma (Burney et al., 1990) has markedly increased. An association between the increase in childhood obesity and asthma has been suggested (Schwartz et al., 1990; Gennuso et al., 1998; Luder et al., 1998). A large recent study has shown that asthma symptoms are associated with BMI but not with the sum of skinfold thickness (Figueroa-Munoz et al., 2001). Cross-sectional studies have been unable to prove what is cause and effect, as asthma may be perceived as a reason for children not to exercise and could potentially result in an increased sedentary lifestyle and increased rates of obesity in this group.

Other respiratory abnormalities have been reported in association with obesity, including a reduction in functional residual capacity (46% of subjects), impairment of diffusion capacity (33%) (Li et al., 2003), a restrictive defect (18%), and obstructive changes (47%) (Mallory et al., 1989).

Sleep-associated breathing disorders occur, particularly obstructive sleep apnea, with abnormal findings on polysomnography in up to 37% of young people (Mallory et al., 1989; Marcus et al., 1996). Neurocognitive effects are common in obese children with obstructive sleep apnea, resulting in deficits in learning, memory, and vocabulary (Rhodes et al., 1995). The potential for this to result in learning difficulties at school is huge, but to date all studies carried out in this area have been relatively small. Pickwickian obesity-hypoventilation syndrome is a rare but dangerous complication consisting of hypoventilation and cor-pulmonale (Taitz, 1983; Riley et al., 1976). It has a high-mortality risk, and aggressive therapy aimed at weight loss is warranted for those obese children with this syndrome.

ORTHOPEDIC DISEASE

Obese children are at risk of orthopedic complications because their cartilage, bones, and unfused epiphyses have not evolved to carry such substantial quantities of excess body weight.

Blount's disease is due to abnormal growth of the medial aspect of the proximal metaphysis of the tibia. The etiology is unknown, and the prevalence is low; however, up to 67% of those presenting are obese, and there is a significant correlation between the degree of tibial deformity and the degree of obesity. These obese patients need an increased number of operations for full correction (Dietz et al., 1982).

Slipped capital femoral epiphysis (SCFE) is known to be associated with obesity. It occurs because of increased weight on the unfused femoral epiphysis; 50 to 70% of patients with SCFE are reported to be obese (Sorenson, 1968; Kelsey et al., 1972; Wilcox et al., 1988). Two thirds of those with bilateral disease present, on average, a year earlier than nonobese children (Loder et al., 1993). Weight loss needs to be addressed aggressively in these particular children as it may prevent both the recurrence of Blount's disease and the unilateral SCFE from becoming bilateral.

GASTROINTESTINAL DISEASES

Fatty infiltration of the liver frequently occurs in obese children (Frelut et al., 1995; Frelut et al., 1996; Tominga et al., 1995) and may be associated with raised liver enzymes (Frelut et al., 1996). The phrase "non-alcoholic fatty liver disease" (NAFLD) has been used to describe the pathological association with obesity. The clinical characteristics include a male predominance, nonspecific abdominal pain, raised aminotransferase, and hypertriglyceridemia. Weight loss can reverse the abnormalities (Vajro et al., 1994). The pathogenesis is unknown, but it is thought that increased free fatty acids

(Wanless and Lentz, 1990) and hyperinsulinemia (Wanless et al., 1989) both contribute. It is not benign, and progression to fatty hepatitis (steatohepatitis), fibrosis, and cirrhosis has been described in children as young as 9 years old (Rashid et al., 2000).

The prevalence of gallstones is increased in obesity due to an increased proportion of cholesterol in the bile, rendering it less soluble and impairing gall bladder emptying (Hendle et al., 1998). Obesity is present in up to 50% of children with gallstones (Crichlow et al., 1972; Halcomb and O'Neill, 1980; Friesen and Roberts, 1989), and conversely, as in adults, cholecystitis may result following rapid weight reduction (Crichlow et al., 1972).

PSEUDOTUMOUR CEREBRI (PTC)

Pseudotumour cerebri (PTC) is a rare disorder of childhood and adolescence; up to 50% of patients with this disorder are obese (Weisberg and Chutorian, 1977). The disease is characterized by raised intracranial pressure and presents with headache, vomiting, and visual problems. Papilloedema is usually present, a sixth nerve palsy is common, and severe visual impairment may occur. It has been hypothesized that the increase in intracranial pressure is secondary to raised intra-abdominal pressure, leading to an increase in intrathoracic pressure (Sugerman et al., 1999). Aggressive treatment of obesity is indicated in this clinical scenario.

PSYCHOLOGICAL PROBLEMS

Obesity is known to be associated with very significant psychological problems in childhood (Reilly et al., 2003). Obese children are often the victims of bullying at school, and preference tests have shown that 10- to 11-year-old children would prefer friends with a wide range of handicaps rather than those who are obese (Richardson et al., 1961). Poor self esteem is one of the areas adversely affected, and Strauss (2000) reported that by 13 to 14 years of age global self esteem scores were lower in obese children, resulting in increased levels of sadness, loneliness, and nervousness and increased likelihood of tobacco and alcohol consumption.

Societal preoccupation with weight and food intake starts at a young age. In a study, of 7- to 13-year olds, almost one half were concerned about their weight and more than one third had already tried to lose weight (Maloney et al., 1989). Abnormal eating patterns are frequent in obese individuals; binge eating disorders occur in approximately 30% of morbidly obese adolescent girls (Berkowitz et al., 1993).

Society does not view overweight children in a positive light, often associating them with laziness. As a consequence, obese children, particularly adolescents, may demonstrate high levels of anxiety and disturbed body image (Kimm et al., 1991). Because of their size, they may be perceived by adults as older than their chronological age, and this can lead to inappropriate expectations and feelings of frustration and failure if they are unable to fulfil expectations (Dietz, 1998).

The psychological impact of obesity in childhood is widespread and varied, and it is an area that needs to be addressed with young people and their families if we are to enable effective and permanent treatment of the condition.

CONSEQUENCES OF CHILDHOOD OBESITY IN ADULTHOOD

The specific effects of childhood obesity on the morbidity and mortality in adulthood have been difficult to elucidate because all of the risks of adult obesity themselves need to be extracted from the equation. Several longitudinal studies have endeavored to do this with follow up many decades later. However, inevitably, these have included a population born and brought up in quite different social circumstances in earlier years and starting when the incidence of obesity was far lower. This must be acknowledged when interpreting any longitudinal results.

Mortality

Must et al. (1992) followed up participants from the Harvard Growth Study 55 years later. They found that those who were overweight as adolescent males, compared with those who were lean, had an increased mortality from all causes, with a relative risk of 1.8 (95% confidence interval 1.2–2.7) and an increased mortality from coronary heart disease, with a relative risk of 2.3 (95% confidence interval 1.4–4.1). Interestingly, females who were overweight as adolescents were not found to have an increased mortality rate.

Mossberg (1989) in a 40-year follow-up study of overweight children in Sweden found that weight for height SDS tended to be associated positively with mortality rate, with an increase in the overweight subjects compared with a reference group (10.9% vs. 6.5%, respectively; males and females were not considered separately in the mortality statistics). Hoffmans et al. (1988) studied the impact of BMI at 18 years of age on mortality 32 years later in male individuals from the Netherlands. Their population was lean, with only 2% having a BMI >25 kg/m^2, but large numbers were studied, and the negative impact of a high BMI only became apparent after 20 years of follow up. Compared with those with a BMI of 19.00 kg/m^2, the risk for men with a higher baseline BMI of 20 to 24 kg/m^2 increased 10%, and those with a BMI of >25 kg/m^2 had an increased mortality risk of 1.57. From these studies, it appears that overweight young people, particularly men, have an increased mortality risk in later life.

Cardiovascular Risks

The cardiovascular system has received more attention than other areas in demonstrating that childhood obesity increases risk factors and then adult morbidity (Sinaiko et al., 1999; Mossberg, 1989; Lauer et al., 1988; Lauer and Clarke, 1989; Srinivasan et al., 1996). However, this may be confounded by the significant element of tracking of childhood obesity into adulthood. Freedman et al. (2001) demonstrated that being overweight as a child is related to adverse cardiovascular risk factors in adults, but associations were weak when adjusted for adult weight. Seventy-seven percent of the overweight children in their sample remained obese as adults, and those that had an earlier onset of obesity, i.e., <8 years, had a higher BMI in adult life than those who became overweight later.

It is clear that obesity, regardless of whether it has its onset in childhood or adulthood, has adverse cardiovascular effects.

Orthopedic Disease

Adolescent overweight has been demonstrated to be associated with increased gout in men and arthritis in women (Must et al., 1992; Lake et al., 1997). Men who were obese as children are up to three times more likely to suffer from gout, whereas women are twice as likely to report arthritis and eight times more likely to report limited mobility than their respective peer groups.

Socioeconomic Issues

Adolescent and early adulthood obesity in females can have a detrimental effect on socioeconomic status during adult life. These women are up to 20% less likely to be married; in the U.S., these women have a reported lower household income (by $6710 a year) and have increased rates of household poverty, independent of their status at baseline and their aptitude score (Gortmaker et al., 1993). These parameters were compared with women who had other chronic conditions, but they did not have the same adverse effects.

The males who were obese as adolescents did not have significant differences, however, when compared with those who had other chronic conditions, but they were 11% less likely to be married.

These data indicate some of the severe long-term negative psychological impacts that adolescent obesity can have on adult life.

CONCLUSION

Childhood obesity clearly has health implications, both during childhood and persisting into adult life. The consequences are multifaceted, affecting many aspects of physiological and psychological functioning.

Further longitudinal studies are needed in this area to continue to delineate cause and effect so that we can try and target an approach to management. Identifying high-risk groups, in whom the consequences of obesity will have a significantly greater impact, may prove useful in the appropriate allocation of resources.

To date, prevention and management of pediatric obesity have been rather piecemeal, and there is minimal good quality evidence on the effectiveness of interventions. Nevertheless, the ever-increasing rise in the prevalence of childhood obesity has grave connotations for the future health of the nation, and it is vitally important that we adopt a multidisciplinary approach to both prevention and treatment of this modern epidemic.

REFERENCES

Adair, L. and Gordon-Larsen, P., Maturational timing and overweight prevalence in US adolescent girls, *Am. J. Public Health* 91, 642–644, 2001.

Aye, T. and Levitsky, L. Type 2 diabetes: an epidemic disease, *Curr. Opin. Pediatr.,* 115, 411–415, 2003.

Berkowitz, R., Stunkard, A., and Stallings, V., Binge-eating disorder in obese adolescent girls, *Ann NY Acad. Sci.,* 699, 200–206, 1993.

Betts, P., Ward, P., Mulligan, J., Smith, B., and Wilkin, T., Increasing body weight predicts the earlier onset of insulin-dependant diabetes in childhood: testing the "accelerator hypothesis" (2), *Diabet. Med.,* 22, 144–151, 2005.

Beunen, G., Malina, R., Lefevre, J., Claessens, A., et al., Adiposity and biological maturity in girls 6–16 years of age, *Int. J. Obes. Relat. Metab. Disord.,* 18, 542–546, 1994.

Burney, P., Chinn, S., and Rona, R., Has the prevalence of asthma increased in children? Evidence from the national study of health and growth 1973–1986, *BMJ* 300, 1306–1310, 1990.

Crichlow, R., Seltzer, M., and Jannetta, P., Cholecystitis in adolescents, *Dig. Dis.,* 17, 68–72, 1972.

Daniels, S., Kimball, T., Morrison, J., Khoury, P., Witt, S., and Meyer, R., The effect of lean body mass, fat mass, blood pressure and sexual maturation on left ventricular mass in children and adolescents: statistical, biological and clinical significance, *Circulation* 92, 3249–3254, 1995.

Daniels, S., Morrison, J., Sprecher, D., Khoury, P., and Kimball, T., Association of body fat distribution and cardiovascular risk factors in children and adolescents, *Circulation* 99, 541–545, 1999.

De Simone, G., Devereux, R., Koren, M., Mensab, G., Casale, P., and Laragh, J., Midwall left ventricular mechanics: an independent predictor of cardiovascular risk in arterial hypertension, *Circulation,* 93, 259–265, 1996.

Despres, J., Lamarche, B., Mauriege, P., et al., Hyperinsulinaemia as an independent risk factor for ischemic heart disease, *N. Engl. J. Med.,* 334, 952–957, 1996.

Dietz, W., Gross, W., and Kirkpatrick, J., Blount disease (tibia vara): another skeletal disorder associated with childhood obesity, *J. Pediatr.,* 101, 735–737, 1982.

Dietz, W., Health consequences of obesity in youth: childhood predictors of adult disease, *Pediatrics,* 101, 518–525, 1998.

Drake, A., Smith, A., Betts, P., Crowne, E., and Sheild, J., Type 2 diabetes in obese white children, *Arch. Dis. Child.*, 86, 207–208, 2002.

Feltbower, R., McKinney, P., Campbell, F., Stephenson, C., and Bodansky, H., Type 2 and other forms of diabetes in 0–30 year olds: a hospital based study in Leeds, UK, *Arch. Dis. Child.*, 88, 676–679, 2003.

Figueroa-Colon, R., Franklin, F., Lee, J., Aldridge, R., and Alexander, L., Prevalence of obesity with increased blood pressure in elementary school-aged children., *South Med. J.*, 90, 806–813, 1997.

Figueroa-Munoz., J.I., Chinn, S., and Rona R., Association between obesity and asthma in 4–11 year old children in the UK, *Thorax*, 56, 133–137, 2001.

Freedman, D., Dietz, W., Srinivasan, S., and Berenson, G., The relation of overweight to cardiovascular risk factors among children and adolescents: The Bogalusa Heart Study, *Pediatrics*, 103, 1175–1182, 1999.

Freedman, D., Kettel Khan, L., Dietz, W., Srinivasan, S., and Berenson G., Relationship of childhood obesity to coronary heart disease risk factors in adulthood: The Bogalusa Heart Study, *Pediatrics*, 108, 712–718, 2001.

Frelut, M., Razakarivony, R., Cathelinau, L., and Navarro, J., Uneven occurrence of fatty liver in morbidly obese children: risk factors and impact of weight loss, *Int. J. Obes.*, 19 (suppl. 2), 122, 1995.

Frelut, M., Wiling, T., Navarro, J., and Debre, R., Fatty liver and hyperaminotransferasemia (ALT) in obese children: reversibility and correlation with fattening pattern but not overweight degree, *Int. J. Obes.*, 20 (suppl. 4), 147, 1996.

Friesen, C. and Roberts, C., Cholelithiasis: clinical characteristics in children, *Clin. Pediatr.*, 7, 294–299, 1989.

Gennuso, J., Epstein, L., Paluch, R., et al., The relationship between asthma and obesity in urban minority children and adolescents, *Arch. Pediatr. Adolescent Med.*, 52, 1197–1200, 1998.

Gortmarker, S., Must, A., and Perrin J., Social and economic consequences of overweight in adolescence and young adulthood, *N. Engl. J. Med.*, 329, 1008–1012, 1993.

Guo, S. and Chumlea, W., Tracking of body mass index in children in relation to overweight in adulthood, *Am. J. Clin. Nutr.*, 70 (suppl.), 145S–148S, 1999.

Gutin, B., Treiber, F., Owens, S., and Mensab, G., Relations of body composition to left ventricular geometry and function in children, *J. Pediatr.*, 132, 1023–1027, 1998.

Halcomb, G. and O'Neill J., Cholecystitis, cholelithiasis and common duct stenosis in children and adolescents, *Ann. Surg.*, 191, 626–635, 1980.

Hendle, H., Hojgaard, L., Anderson, T., Pedersen, B., Paloheimo, L., Rehfeld, J., et al., Fasting gall bladder volume and lithogenicity in relation to glucose tolerance, total and intra-abdominal fat masses in obese non-diabetic subjects, *Int. J. Obes. Relat. Metab. Disord.*, 22, 294–302, 1998.

Hoffmans, M., Kromhout, D., and De Lezenne Coulander, C., The impact of body mass index of 78,612 18-year old Dutch men on 32-year mortality from all causes, *J. Clin. Epid.*, 41, 749–756, 1988.

Hubert, H., Feinleib, M., McNamara, P., and Castelli, W., Obesity as an independent risk factor for cardiovascular disease: a 26-year follow-up of participants in the Framlingham Heart Study, *Circulation*, 67, 968–977, 1983.

Kannel, W., D'Aostino, R., and Cobb, J., Effect of weight on cardiovascular disease, *Am. J. Clin. Nutr.*, 63 (suppl. 1), 419S–422S, 1996.

Kaplowitz, P., Slora, E., Wasserman, R., Pedlow, S., and Herman-Giddens, M., Earlier onset of puberty in girls: relation to increased body mass index and race, *Pediatrics,* 108, 347–353, 2001.

Kelsey, J., Acheson, R., and Keggi, K., The body build of patients with slipped femoral capital epiphysis, *Am. J. Dis. Child.,* 124, 276–281, 1972.

Kibirige, M., Metcalf, B., Renuka, R., and Wilkin, T., Testing the Accelerator Hypothesis: the relationship between body mass and age at diagnosis of type 1 diabetes, *Diabetes Care,* 26, 2865–2870, 2003.

Kimm, S., Sweeney, C., and Janosky J., Self-concept measures and childhood obesity: a descriptive analysis, *J. Dev. Behav. Pediatr.,* 12, 19–24, 1991.

Lake J., Power C., and Cole T., Women's reproductive health: the role of body mass index in early and adult life, *Int. J. Obes.,* 21, 432–438, 1997.

Lauer, R. and Clarke, W., Childhood risk factors for high adult blood pressure: The Muscatine Study, *Pediatrics,* 84, 633–641, 1989.

Lauer, R., Lee, J., and Clarke W., Factors affecting the relationship between childhood and adult cholesterol levels: The Muscatine Study, *Pediatrics,* 82, 309–318, 1988.

Li, A., Chan, D., Wong, E., Yin, J., Nelson, E., and Fok, T., The effects of obesity on pulmonary function, *Arch. Dis. Child.,* 88, 361–363, 2003.

Loder, R., Arbor, A., Aronson, D., and Greenfield M., The epidemiology of bilateral slipped capital femoral epiphysis, *J. Bone Joint Surg.,* 75, 1141–1147, 1993.

Luder, E., Melnik, T., and DiMaio, M., Association of being overweight with greater asthma symptoms in inner city black and Hispanic children, *J. Pediatr.,* 132, 699–703, 1998.

Malecka-Tandera, E., Wrzesniewski, N., Kurkowska, M., and Kudla, M., Overweight in adolescent girls with menstrual irregularities is a risk factor for polycystic ovary syndrome, *Int. J. Obes.,* 22 (suppl. 4), S26, 1998.

Mallory, G., Fiser, D., and Jackson, R., Sleep-associated breathing disorders in morbidly obese children and adolescents, *J. Pediatr.,* 115, 892–897, 1989.

Maloney, M., McGuire, J., Daniels, S., and Specker, B., Dieting behaviour and eating attitudes in children, *Pediatrics,* 84, 482–489, 1989.

Marcus, C., Curtis, S., Koerner, C., Joffe, A., Serwint, J., and Loughlin, G., Evaluation of pulmonary function and polysomnography in obese children and adolescents, *Pediatr. Pulmonol.,* 21, 176–183, 1996.

McCarthy, H., Ellis, S., and Cole, T., Central overweight and obesity in British youth aged 11–16 years: cross sectional surveys of waist circumference, *BMJ,* 326, 624, 2003.

Mossberg, H., 40 year follow up of overweight children, *Lancet* 2, 491–193, 1989.

Must, A., Jacques, P., Dallal, G., Bajema, C., and Dietz, W., Long-term morbidity and mortality of overweight adolescents, *N. Engl. J. Med.,* 327, 1350–1355, 1992.

Must, A., Spadano, J., Coakley, E., Field,. A., Colditz, G., and Dietz W., The disease burden associated with overweight and obesity, *JAMA,* 282, 1523–1529, 1999.

Pinhas-Hamiel, O., Dolan, L., Daniels, S., Standiford, D., Khoury, P., and Zeitler P., Increased incidence of on-insulin-dependent diabetes mellitus among adolescents, *J. Pediatr.* 128, 608–615, 1996.

Power, C., Lake, J., and Cole, T., Body mass index and height from childhood to adulthood in the 1958 British born cohort, *Am. J. Clin. Nutr.,* 66, 1094–1101, 1997.

Rashid, M. and Roberts, E., Non-alcoholic steatohepatitis in children, *J. Pediatr. Gastro. Nutr.,* 30, 48–53, 2000.

Reilly, J., Methven, E., Hacking, B., Alexander, D., Stewart, L., and Kelnar, C., Health consequences of obesity, *Arch. Dis. Child.,* 88, 748–752, 2003.

Rexrode. K., Manson, J., and Hennekens, C., Obesity and cardiovascular disease, *Curr. Opin. Cardiol.*, 11, 490–495, 1996.

Rhodes, S., Shimoda, K., Waid, L., et al., Neurocognitive deficits in morbidly obese children with obstructive sleep apnoea, *J. Pediatr.*, 127, 741–744, 1995.

Richardson, S., Goodman, N., Hastorf, A., and Dornbusch, S., Cultural uniformity in reaction to physical disabilities, *Am. Soc. Rev.*, 26, 241–247, 1961.

Riley, D., Santiago, T., and Edelman, N., Complications of obesity-hypoventilation syndrome in childhood, *Am. J. Dis. Child.*, 130, 671–674, 1976.

Rosenfield, R., Hyperandrogenism in peripubertal girls, in *Current Issues in Pediatric and Adolescent Endocrinology,* Mahoney, C.P., Ed., Philadelphia, WB Saunders Co., pp. 1333–1358, 1990.

Rudolf, M., Sahota, P., Barth, J., and Walker J., Increasing prevalence of obesity in primary school children: cohort study, *BMJ* 322, 1094–1095,, 2001.

Schwartz, J., Gold, D., Dockery, D., et al., Predictors of asthma and persistent wheeze in a national sample of children in the United States, *Am. Rev. Respir. Dis.*, 142, 555–562, 1990.

Serdula, M., Ivery, D., Coates, R., Freedman, D., Williamson, D., and Byers, T., Do obese children become obese adults? A review of the literature, *Prev. Med.*, 22, 167–177.

Sinaiko, A., Donahue, R., Jacobs, D., and Prineas, R., Relation of weight and rate of increase in weight during childhood and adolescence to body size, blood pressure, fasting insulin, and lipids in young adults, *Circulation,* 99, 1471–1476, 1999.

Sinha, R., Fisch, G., Teague, B., Tamborlane, W.V., Banyas, B., Allen, K., Savoye, M., Rieger, V., Taksali, S., Barbetta, G., Sherwin. R.S., and Caprio S., Prevalence of impaired glucose tolerance among children and adolescents with marked obesity, *N. Engl. J. Med.*, 346, 802–10, 2002.

Sorenson, K., Slipped upper femoral epiphysis, *Acta Orthop. Scand.*, 39, 499–517, 1968.

Srinivasan, S., Bao, W., Wattigney, W., and Berenson, G. Adolescent overweight is associated with adult overweight and related multiple cardiovascular risk factors: The Bogalusa Heart Study, *Metabolism* 45, 235–240, 1996.

Strauss, R., Childhood obesity and self-esteem. *Pediatrics,* 105, E15, 2000.

Sugerman, H., Felton, W., and Sismani, A., Gastric surgery for pseudotumour cerebri associated with severe obesity, *Ann. Surg.*, 229, 634–640, 1999.

Taitz, L., Prognosis of the obese child, in *The Obese Child,* Taitz, L. and Leonard, S., Eds., Oxford, Blackwell Scientific Publications pp. 166–170, 1983.

Tominaga, K, Kurata, J., Chen, Y., et al., Prevalence of fatty liver in Japanese children and relationship to obesity. An epidemiological ultrasonographic survey, *Dig. Dis. Sci.*, 40, 2002–2009, 1995.

Vajro, P., Fontanella, A., and Perna, C., Persistent hyperaminotransferasemia resolving after weight reduction in obese children, *J. Pediatr.*, 125, 239–241, 1994.

Voors, A., Harsha, D., Webber, L., and Berenson, G., Obesity and external sexual maturation — the Bogalusa Heart Study, *Prev. Med.*, 10, 50–61, 1981.

Wabitsch, M., Hauner, H., Heinze, E., Muche, R., Bockmann, A., et al., Body fat distribution and steroid hormone concentrations in obese adolescent girls before and after weight reduction, *J. Clin. Endo. Metab.*, 80, 3469–3475, 1995.

Wang, Y., Is obesity associated with early sexual maturation? A comparison of the association in American boys versus girls, *Pediatrics,* 110, 903–910, 2002.

Wanless, I., Bargman, J., Oreopoulos, D., and Vas, S., Subcapsular steatonecrosis in response to peritoneal insulin delivery: a clue to the pathogeesis of steatonecrosis in obesity, *Modern Pathol.*, 2, 69–74, 1989.

Wanless, I. and Lentz, J., Fatty liver hepatitis (steatohepatitis) and obesity: an autopsy study with analysis of risk factors, *Hepatology* 12, 1106–1110, 1990.

Weisberg, L. and Chutorian A., Pseudotumour cerebri of childhood, *Am. J. Dis. Child.* 131, 1243–1248, 1977.

Whincup, P., Glig, J., Papacosta, O., Seymour, C., Miller, G., Alberti, K., and Cook, D., Early evidence of ethnic differences in cardiovascular risk: cross sectional comparison of British South Asian and white children, *BMJ*, 324, 635–638, 2002.

Whitaker, R., Wright, J., Pepe, M., Seidel, K., and Dietz, W., Predicting obesity in young adulthood from childhood and parental obesity, *N. Engl. J. Med.*, 337, 869–873, 1997.

Wilcox, P., Weiner, D., and Leighley, B., Maturation factors in slipped capital femoral epiphyses, *J. Pediatr. Ortho.*, 8, 196–200, 1988.

Wilkin, T., The accelerator hypothesis: weight gain as the missing link between type 1 and type 2 diabetes, *Diabetologia*, 44, 914–922, 2001.

World Health Organization, Obesity: preventing and managing the global epidemic, Geneva: WHO, 1998.

4

SOCIAL AND SELF-PERCEPTION OF OBESE CHILDREN AND ADOLESCENTS

Andrew Hill

CONTENTS

INTRODUCTION

There is a burgeoning interest in psychosocial issues related to child obesity. It is likely that this stems from the well-publicized impact of obesity on physical health, the need to implement preventative actions, and the identification of childhood as a place to start. This interest also acknowledges that children are in need of protection from, among other things, the distress associated with being obese. The body weight and shape dissatisfaction arising from child and adolescent overweight is well documented (e.g., Wadden et al., 1989). The perspective on distress is broadening to include depression (Erickson et al., 2000), psychiatric disorders (Lamertz et al., 2002, Mustillo et al., 2003), and quality of life (Schwimmer et al., 2003).

In contrast, however, the association between low self-esteem and obesity has a much longer history. So far, the research outcome has not been clear, being compromised by several issues including self-esteem conceptualization and assessment. In this chapter, some of the recent developments in self-esteem research are considered. In particular, I will consider how obese children view themselves and in what ways this differs from nonobese children. Attention will be given to aspects of their social world and whether achieving weight loss has an impact on their social and self-perception. I will argue that there is variability in the relationship between child obesity and self-esteem. Age, gender, and victimization experiences are strong determinants, and their impact is often competence specific.

OBESITY AND GLOBAL SELF-ESTEEM

Two reviews have considered the relationship between obesity and self-esteem. Looking specifically at the literature on children and adolescents, French et al. (1995) found that only 13 of the 25 cross-sectional studies reviewed showed significantly lower self-esteem in obese youngsters. Studies with older children (13 to 18 years) revealed the clearest associations, but little difference was apparent in those aged 7 to 12. Miller and Downey (1999) conducted a meta-analysis of all publications that included a global measure of self-esteem and provided the relevant data for testing the relationship between overweight and self-esteem, regardless of age of participant. In all, they found 91 statistically significant associations reported in 71 studies. The mean weighted correlation between overweight and global self-esteem was 0.18, equivalent to Cohen's d = 0.36 [95% confidence interval (CI) 0.33 to 0.40]. Overall, this indicated a robust but small to moderate size relationship between the two variables.

Important influences on the strength of this relationship were age and gender. Miller and Downey (1999) found that the correlation between overweight and self-esteem increased with increasing age, from 0.12 to 0.22

to 0.28 in children, adolescents, and young adults, respectively. In addition, the relationship was significantly stronger in females (0.23) than in males (0.09).

A Weak Relationship

But why is this a relatively weak relationship? First, many factors contribute to the development and maintenance of obesity. Low self-esteem is probably a minor contributor, albeit one with the potential to interact with other risk factors. And overweight is undoubtedly only one of several influences on an individual's sense of self-worth. Second, both are resistant to change, and correlational relationships are stronger when change in one variable can be associated with change in another. Obesity is notoriously difficult to reduce, and even when weight is lost much of it is regained. Self-esteem is also resistant to change because it is a higher-order schema or internalized theory about the self. Any threat to stability produces anxiety and is therefore avoided. It comes as a surprise to many that low self-esteem, like high self-esteem, is strongly defended. Third, there has been insufficient attention to the conceptualization and measurement of self-esteem. Many psychologists would argue that, at its most basic, self-esteem represents a generalized feeling about the self that is more or less positive (Emler, 2001). This evaluation of global self-worth was the focus of Miller and Downey's (1999) review. However, there are more elaborated conceptualizations, two of which are outlined below.

PERCEIVED SELF-COMPETENCE

The representation of self-esteem as the ratio of a person's successes to their pretensions has been attributed to William James. From this perspective, self-esteem is a personal evaluation of competence in areas that are deemed personally and culturally important. In addition, the same person may consider the importance of domains in which they should feel competent. It is this discrepancy between competence and importance that defines overall self-worth. Only when a person feels low competence in an area of high importance is their self-worth jeopardized.

Susan Harter developed this conceptualization together with a series of measures of perceived self-competence for different age groups (Harter, 1993). She argues that, for children, domains of competence are set by parents (scholastic competence, behavioural conduct) and peers (physical appearance, social and athletic competence) and that these expand in number and range through adolescence into adulthood. In the questionnaire assessments, Harter included several items from each domain, with each item presenting respondents with two contrasting statements; e.g., "Some kids find it *hard* to make friends BUT Other kids find it pretty *easy* to make friends." The children first chose which description was most like them and then rated their choice as "Sort of true for me" or "Really true for me." The assessment also included the children rating how important they feel it is to be successful in each area. The result is a profile of self-rated competence in several domains, an assessment of global self-worth, together with a measure of the perceived importance of each domain.

Research in Obese Children

A few years ago, a colleague and I reported that a community sample of obese 9-year-old girls scored significantly lower on physical appearance and athletic competence than their normal-weight peers (Phillips and Hill, 1998). Their mean scores fell just below the midpoint of these scales, indicating that, on average, these obese girls identified with the depictions of an unattractive and unathletic child. Although global self-worth was lower than that of normal-weight girls, the difference just failed to reach significance.

Older girls show a similar impact of obesity. In a further study, obese 12-year olds had significantly lower athletic competence and global self-worth and marginally lower physical appearance self-esteem (Murphy and Hill, in preparation). Interestingly, there was no association in either the 12- or 9-year olds between obesity and domain importance, suggesting that the obese girls were not responding to their reduced sense of competence by devaluing the importance of success in these domains.

Two additional observations are of note from research using this methodology. First, the impact of overweight on girls' self-competence may be detectable as early as at age 5 years (Davison and Birch, 2001). Perceived cognitive ability was significantly lower than that of normal-weight girls but was still relatively high. Second, boys show some resistance to these relationships, at least while they are young. We found no relationship between obesity and low self-esteem on any aspect of perceived self-competence in 9- and 12-year-old boys (Hill et al., in preparation). Strauss (2000) reported that it was only in 13- to 14-year-old boys that a small but significant decrease in global self-worth became apparent. Similarly, French et al. (1996) found obese teenage boys to have reduced competence in athletic ability, physical appearance, and romantic appeal.

THE SOCIAL BASIS OF SELF-ESTEEM

An alternative perspective on self-esteem derives from Charles Cooley's notion of the looking glass self (Emler, 2001). In this perspective, self-esteem is regarded as primarily social in nature and based on judgments that we imagine others to make of us. Those with low self-esteem perceive others to have little regard for them and feel demeaned, neglected, or socially isolated. Especially influential are significant others (such as parents for children) and successful people who we imagine judge us more harshly than those less successful or virtuous.

Peer Nominations

So what does this perspective hold for obesity? None of the studies cited above using Harter's measure with preteenagers showed any weight-related difference in perceived social competence. This positive social view has some external credibility. For example, we have used a peer nomination procedure to investigate children's popularity with their same sex peers in the classroom (e.g., Phillips and Hill, 1998). This required children to identify up to three others in their school class group that they would choose to sit next to in class, play with at break time, and invite home to tea. Neither 9- nor 12-year-old obese girls or boys were disadvantaged in terms of peer choices. They were as likely to be chosen as their lean peers as people to socialize with inside and outside of school.

The one caveat is that a similar evaluation of attractiveness showed that obese girls were much less likely to be identified as attractive. Because attractiveness was the most important predictor of peer-rated popularity, it may only be a matter of time before obese adolescent girls suffer a decline in popularity. Indeed, this was apparent in a study of U.S. 13- to 18-year olds. This study used a peer-nomination procedure similar to that described above and found that the overweight and obese adolescents were more isolated and peripheral to social networks than their leaner counterparts (Strauss and Pollack, 2003). Although they named a similar number of friends as their normal-weight peers, overweight adolescents received significantly fewer friendship nominations by others. It appears that their outward show of affiliation was not reciprocated.

Sociometer Theory

Mark Leary has extended the social view of self-esteem to question its basic function (Leary, 1999). Sociometer theory proposes that the self-esteem system evolved primarily as a monitor of social acceptance, the motivation being not to maintain self-esteem per se but to avoid social devaluation and rejection. Leary argues that people are particularly sensitive to changes in relational evaluation or the degree to which others regard their relationship with the individual as valuable, important, or close. Accordingly, self-esteem is lowered by failure, criticism, or rejection and raised by success, praise, and events associated with relational appreciation. Even the possibility of rejection can lower self-esteem.

The Impact of Victimization

Taking Leary's perspective on self-esteem, it is pertinent to look at circumstances where people have been socially rejected by being victimized because of their overweight. Some 40% of obese adults in the U.S. report having experienced weight-related victimization (Falkner et al., 1999). The most frequently encountered stigmatizing situations include comments from children, people making negative personal assumptions, being stared at, and inappropriate comments from doctors (Myers and Rosen, 1999). Between one quarter and one third of adolescent teenagers report being teased by peers for reasons of weight (Eisenberg et al., 2003). Interestingly, there was a gender-by-weight interaction in that obese girls and thin boys reported the highest levels of teasing. Regardless, in both of these studies, victimization was associated with psychological distress. Those teased about their weight had low body satisfaction, low self-esteem, higher levels of depressive symptoms, and suicidal ideation, even after controlling for differences in body weight.

Weight-related victimization fails to distinguish victimization for fatness from that for thinness. Our own research has looked specifically at fat-teasing — being teased, bullied, or called horrible names for being fat — using questionnaire items inserted into Harter's Self-Perception Profile for Children. In one study of around 450 12-year olds, 12% of girls and 16% of boys identified with the description of themselves as a fat-teased child (Hill and Murphy, 2000). Although these children were heavier than their non-victimized peers, less than one half were either overweight or obese. Again, being fat-teased was associated with low body shape satisfaction and low self-esteem. As Figure 4.1 shows, fat-teased girls and boys had significantly

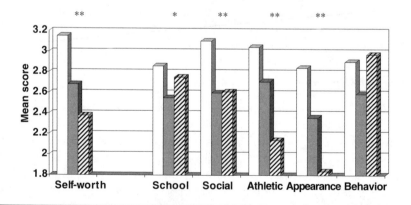

Figure 4.1   Self-Esteem in 12-Year-Old Children Not Victimized (Open Columns), Fat-Teased Boys (Closed Columns), and Fat-Teased Girls (Hatched Columns). [Data From Hill and Murphy, *Int. J. Obes.*, 24 (Suppl. 1), 161, 2000.] Significant Impact of Victimization, *p < .05 and **p < .001.

lower global self-worth and lower perceived competence in all domains except behavioral conduct. Fat-teased girls in particular saw themselves as unattractive, unathletic, and with low global self-worth. These differences remained even after controlling for differences in body weight.

A similar study of nearly 400 9-year olds showed that 21% of girls and 16% of boys were fat-teased (Hill and Waterston, 2002). Like the older group, fat-teased children scored significantly lower on global self-worth and all measures of self-competence. Overweight and obese children were more likely to be victimized. The relative risk of being fat-teased (compared with normal-weight children) was 3.9 (95% CI 2.6 to 5.7). Again, only half of those who were overweight endorsed the description of a fat-teased child. Indeed, half of those victimized were statistically within normal weight boundaries.

Two final points are noteworthy. First, although being fat-teased was much more common in overweight and obese children, at least half did not report these experiences. We know very little about what protects these children or what makes the other half vulnerable. Second, in neither study did victimization have an impact on the perceived importance of any of these domains. Once more, it would appear that these children were not managing their low self-esteem by modifying the importance of domains in which they judged themselves less competent.

LOSING WEIGHT

If child (and adult) obesity does impact on self-esteem, then treatments that lead to weight loss should be accompanied by self-esteem improvements. Moreover, in an ideal world, the two outcomes should be related to each other. Unfortunately, in the intervention studies reviewed by French et al. (1995), an increase in self-esteem was observed in a few studies but was never correlated with any decrease in body weight. One reason may be the general failure to bring about appreciable weight loss.

Self-Esteem Changes

In another study, self-esteem in obese adolescents attending a summer residential weight-loss camp in the U.K. was examined (Walker et al., 2003). Staying for an average of 4 weeks, the 57 children included in the report (mean age 14 years) lost 5.6 kg and reduced their BMI by 2.1 kg/m^2 and BMI SD score by 0.28. Comparison children gained weight by a small but statistically significant amount.

Figure 4.2 summarizes the self-esteem of obese adolescents at the start and end of the camp. Overall, their scores were significantly lower than the comparison adolescents on every measure other than social acceptance. Significant group-by-time interactions showed that, by the end of their stay, the obese adolescents had increased their global self-worth, physical

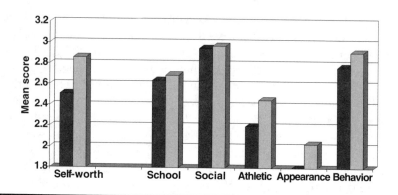

Figure 4.2 Self-Esteem in Adolescents Attending a Summer Weight-Loss Camp at the Start (Dark Columns) and End (Light Columns) of their Stay. [Data from Walker et al., *Int. J. Obes.*, 27, 748–754, 2003.] Significant Pre-post Differences, **p < .001.

appearance, and athletic competence. However, despite the improvements, they still fell short of the positive self-perception of the normal weight comparison group. It is of note that the degree of BMI loss was correlated with improvements in self-worth ($r = 0.50$, $p < .001$), physical appearance ($r = 0.39$, $p < .001$), and athletic competence ($r = 0.28$, $p < .01$). BMI was not associated with domain importance either at the start or end of the camp. Indeed, in neither group of adolescents did domain importance change over the assessment period. The main psychological achievement of these weight-reducing obese teenagers was to improve their current perception of physical attractiveness and athletic ability rather than renegotiate the perceived importance of these attributes.

Cognitive Changes

Looking in more detail at the process of psychological change, we used a sentence completion methodology to evaluate the cognitions of obese adolescents attending the weight-loss camp (Barton et al., 2004). The focus of this study was the automatic thoughts and conditional beliefs these teenagers had about exercise, eating, and appearance. Automatic thoughts are situation-specific unplanned thoughts that can be positive, neutral, or negative in tone. Conditional beliefs are underlying assumptions about the consequences of hypothetical situations, and range from functional, adaptive responses to those that are dysfunctional and maladaptive. By providing respondents with sentence stems such as "Physical activity makes me feel ..." or "If I chose to over-eat ..." and asking for the sentence to be completed in their own words, it is possible to explore people's self-representations in these areas. We found that the obese adolescents had significantly more dysfunctional conditional beliefs about exercise, eating, and appearance compared with normal-weight teenagers and that these did not change over the course of the camp. In other words, at a schema level, these adolescents saw themselves as dealing with exercise, eating, and appearance situations in a more negative and maladaptive manner. This view was difficult to change.

In contrast, although they also had more negative and fewer positive automatic thoughts about these issues than lean comparison teenagers, this more superficial level of self-representation did change over the camp. By the end of camp, automatic thoughts about exercise and appearance became less negative and more positive, whereas those about eating did not change. For example, extreme negative completions for the sentence stem "I feel that my shape ..." had reduced from 71% to 34% of the adolescents. Completions such as "is horrible," "is too big," or "is disgusting" were replaced by more positive completions, including "is better," "nicer," "OK," or "alright." Although these did not reach the positive self-representation shown by lean comparisons, the improvements in cognitive content were correlated with degree of weight change. Overall, these studies show that success in weight loss can bring about important changes in self-representation and esteem.

CONCLUSIONS

In order to appreciate how obese youngsters feel about themselves, it is valuable to use assessment approaches derived from theories about self-esteem. Research using Harter's perceived self-competence measure suggests that obese children do not view themselves as being without merit. Although global self-worth may be reduced, they see themselves as well-behaved, good at school work, and as having friends, at levels similar to their lean peers (e.g., Phillips and Hill, 1998). It is their physical appearance and athletic competence that are most greatly influenced. Age and gender strongly determine this self-perception. For girls, these associations are detectable before puberty. For boys, it is only during early teenage years that self-competence is impaired. As teenage years advance, the range of areas of competence becomes increasingly influenced by their obesity.

Social interaction is key to self-perception. For younger children, their weight has little impact on their perceived or actual friendship status. By teenage years, peer relationships become problematic, with obese adolescents being more at risk of marginalization and outright victimization. Presumably, this reflects the different basis on which friendship networks are organized as children mature and as dating and sexual relationships commence. Those obese children victimized for their obesity have low self-esteem, which in some areas is crushingly low. There is still little definitive evidence on causation. While it may appear obvious that victimization lowers self-esteem, it is also the case that those with low self-esteem treat themselves badly and invite bad treatment from others (Emler, 2001). A sustaining cycle of low self-esteem and social rejection is extremely plausible.

It is important to recognize that this analysis derives mainly from community samples and that the description is extremely individually variable. Clinical studies of obese youngsters tend to reveal more psychological dysfunction than do community selected groups, a sampling bias that is common in the psychological literature. Additionally, a significant proportion of obese children appear protected from, or resistant to, the psychological consequences of their obesity. It is unclear whether they will remain so as adults. However, this group are a potentially important source of information regarding how we might assist in the rehabilitation of those deeply affected. For example, is there something about their parenting, peer group, or interpersonal style that could form the basis for a psychological intervention to improve self-representation? This variability also means that we as health professionals should not go looking for psychological distress in every obese child. And even if we find it, we should question whether its cause lies in their obesity or is determined by something else.

Improving Self-Esteem?

This chapter has paid close attention to some of the psychological processes in self-representation and the maintenance of self-esteem. Harter has suggested that low self-esteem adolescents may seek to improve their situation by discounting the importance of domains in which they perceive themselves to be low in competence (Harter, 1993). No evidence for this was apparent in our research. However, we did find evidence of an alternative strategy — that of raising perceived competence.

Increased competence in athletic ability and physical appearance seen over the course of the summer weight-loss camps was significantly correlated with amount of weight lost. It is at least as likely that the fun and safe social environment together with the opportunity to engage in a range of physical activities was as necessary as the weight loss for these improvements. Lowering a person's aspirations by discounting the importance of, for example, physical appearance may be futile in the face of the high societal valuation of external appearance. More feasible for interventions to improve self-esteem is the raising of self-competence in specific areas. The weight-loss camp findings suggest the benefit of repeated or sustained positive experiences rather than one-off events in order to raise self-competence. These findings also argue for attention to psychological issues in the management of child obesity and their assessment in the evaluation of treatment outcome.

REFERENCES

Barton, S.B., Walker, L.L.M., Lambert, G., Gately, P.J., and Hill, A.J., Cognitive change in obese adolescents losing weight, *Obes. Res.,* 12, 313–319, 2004.

Davison, K.K. and Birch, L.L., Weight status, parent reaction, and self-concept in five-year-old girls, *Pediatrics,* 107, 46–53, 2001.

Eisenberg, M.E., Neumark-Sztainer, D., and Story, M., Associations of weight-based teasing and emotional well-being among adolescents, *Arch. Pediatr. Adolesc. Med.,* 157, 733–788, 2003.

Emler, N., *Self-Esteem: The Costs and Causes of Low Self-Worth,* Joseph Rowntree Foundation, York, 2001.

Erickson, S.J., Robinson, T.N., Haydel, K.F., and Killen, J.D., Are overweight children unhappy? Body mass index, depressive symptoms, and overweight concerns in elementary school children, *Arch. Pediatr. Adolesc. Med.,* 154, 931–935, 2000.

Falkner, N.H., French, S.A., Jeffery, R.W., et al., Mistreatment due to weight: prevalence and sources of perceived mistreatment in women and men, *Obes. Res.,* 7, 572–576, 1999.

French, S.A., Perry, C.L., Leon, G.R., et al., Self-esteem and changes in body mass over 3 years in a cohort of adolescents, *Obes. Res.,* 4, 27–33, 1996.

French, S.A., Story, M., and Perry, C.L., Self-esteem and obesity in children and adolescents: a literature review, *Obes. Res.,* 3, 479–490, 1995.

Harter, S., Causes and consequences of low self-esteem in children and adolescents, in *Self-Esteem: The Puzzle of Low Self-Regard,* Baumeister, R.F., Ed., Plenum, New York, 1993, chap. 5.

Hill, A.J. and Murphy, J.A., The psycho-social consequences of fat-teasing in young adolescent children, *Int. J. Obes.,* 24 (Suppl. 1), 161, 2000.

Hill, A.J. and Waterston, C.L., Fat-teasing in pre-adolescent children: The bullied and the bullies, *Int. J. Obes.,* 26 (Suppl. 1), 20, 2002.

Lamertz, C.M., Jacobi, C., Yassouridis, A., et al., Are obese adolescents and young adults at risk for mental disorders? A community survey, *Obes. Res.,* 10, 1152–1160, 2002.

Leary, M., Making sense of self-esteem, *Curr. Directions Psychol. Sci.,* 8, 32, 1999.

Miller, C.T. and Downey, K.T., A meta-analysis of heavyweight and self-esteem, *Personality Soc. Psychol. Rev.,* 3, 68, 1999.

Mustillo, S., Worthman, C., Erkanli, A., et al., Obesity and psychiatric disorder: developmental trajectories, *Pediatrics,* 111, 851–859, 2003.

Myers, A. and Rosen, J.C., Obesity stigmatization and coping: relation to mental health symptoms, body image, and self-esteem, *Int. J. Obes.,* 23, 221–230, 1999.

Phillips, R.G. and Hill, A.J., Fat, plain, but not friendless: self-esteem and peer acceptance of obese pre-adolescent girls, *Int. J. Obes.,* 22, 287–293, 1998.

Schwimmer, J.B., Burwinkle, T.M., and Varni, J.W., Health-related quality of life of severely obese children and adolescents, *JAMA,* 289, 1813–1819, 2003.

Strauss, R.S., Childhood obesity and self-esteem, *Pediatrics,* 105, e15, 2000.

Strauss, R.S. and Pollack, H.A., Social marginalization of overweight children, *Arch. Pediatr. Adolesc. Med.,* 157, 746–752, 2003.

Wadden, T.A., Foster, G.D., Stunkard, A.J. and Linowitz, J.R., Dissatisfaction with weight and figure in obese girls: discontent but not depression, *Int. J. Obes.,* 13, 89–97, 1989.

Walker, L.L., Gately, P.J., Bewick, B.M. and Hill, A.J., Children's weight-loss camps: psychological benefit or jeopardy?, *Int. J. Obes.,* 27, 748–754, 2003.

5

SELF-PERCEPTIONS AND PHYSICAL ACTIVITY BEHAVIOR OF OBESE YOUNG PEOPLE

Lucy Foster and Angie Page

CONTENTS

INTRODUCTION

Enhancing physical activity through increasing active behaviors or decreasing sedentary behaviors offers a promising approach for the treatment of childhood and adolescent obesity (Glenny and O'Meara, 1997; Barlow and Dietz, 1998; Epstein et al., 1998; Story, 1999; Zwiauer, 2000; Summerbell et al., 2001; Wilson et al., 2003). Physical activity can assist not only in short- and long-term weight management but may also have a number of other physiological and psychological benefits for the young person (Fox, 2000). Physical activity behaviors, in addition to eating behaviors and environmental, psychological, familial, and socio-demographic factors, are key considerations when tailoring obesity treatment programs for individual needs (Edmunds et al., 2001). However, for many obese young people, adopting a physically active lifestyle can be hindered by a number of practical, physiological, psychological, and social circumstances.

Obese children and adolescents are reported to be less active than their nonobese peers (Delany et al., 1995; Maffeis et al., 1996; Molnar, 2000; Trost et al., 2001). However, any attempt to change behavior is conditional upon understanding the physical activity experiences of obese young people, how they are determined, and what consequences they have for future involvement. Psychological factors, such as attitudes and beliefs, motivation, self-efficacy, and other physical self-perceptions, play an important role when individuals decide to participate in, and maintain, activity behavior (Biddle and Fox, 1998). Evidence of specific barriers to exercise experienced by obese young people is limited. However, low levels of physical activity self-efficacy, low parental and social support/encouragement, and physical self-worth-related barriers have been previously reported (Trost et al., 2001; Zabinski et al., 2003).

Self-esteem plays a fundamental role in human cognition, motivation, emotion, and behavior (Campbell and Lavallee, 1993). It also is a key driver for participation in physical activity behavior, with physical self-perceptions, in particular, thought to be important for motivation and involvement in physical activity and sport (Biddle and Armstrong, 1992; Crocker et al., 2000). Self-perceptions of body attractiveness, physical competence, physical strength, and physical condition formulate our overall physical self-worth, a key component of self-esteem (Carless and Fox, 2003). Generally, there is a weak but positive association between low self-esteem and obesity in obese young people (French et al., 1995; Miller and Downey, 1999), with the relationship stronger for females rather than males and in older vs. younger populations. However, a stronger inverse relationship does exists between appearance/body-esteem and obesity in young people (Phillips and Hill, 1998; Young-Hyman et al., 2003). Athletic (physical) competence is also reported to be lower in overweight young people compared with nonoverweight peers (French et al., 1996; Phillips and Hill, 1998; Burrows and Cooper, 2002).

Furthermore, obesity treatment studies report the potential for increases in physical self-perceptions of this population (Jelalian and Mehlenbeck, 2002) and that improvement in self-esteem indicators is related to degree of weight loss (Walker et al., 2003).

Thus, on the one hand, we know that self-esteem, particularly physical self-perception, is low in obese young people and that this is often combined with low levels of physical activity, particularly in adolescent girls. Yet, on the other hand, there is evidence that self-worth can be improved by participation in physical activity programs and that increasing self-worth is likely to facilitate further activity involvement. However, there is a view also that physical activity opportunities, particularly in schools, can provide a platform for eroding the self-worth of obese young people. For example, the clothing and overt display of physical competence (or lack thereof) can provide a target for victimization and bullying. Qualitative studies offer a unique opportunity to explore this complex representation of the physical self and how this relates to physical activity behavior. To date, this approach has rarely been used with obese young people (Fox and Edmunds, 2000).

This chapter will draw on a range of interviews conducted with severely obese children and adolescents and their families over the course of 4 years. Specifically, we will seek to address the following questions:

- What are the physical activity experiences of obese young people?
- How is the physical self related to physical activity behavior?
- What approaches can be used to enhance self-perceptions and weight management?

METHODS

All participants were recruited from an outpatient obesity clinic in the South West of England. The school-aged (4 to 18 years) children were obese, as defined by international age- and gender-specific body-mass index (BMI) classification standards for children and adolescents (Cole et al., 2000). All families had been referred from their general practitioner or pediatrician. Appointment times ranged from 2 to 9 mo. Treatment typically involved medical examination (including blood tests) and consultation by a pediatrician and lifestyle advice related to diet and physical activity by a pediatric dietician. Appointments typically lasted from 15 to 30 min.

Thirty-two parents and their children were invited to take part in an interview, of which 24 provided consent and participated in semi-structured interviews. These were conducted with both the young people and their parents while within a quiet and private setting. Interviews aimed to explore their views on obesity referral and treatment processes. Interviews (30 to 60 min) were audio-recorded and transcribed verbatim. Transcripts were imported into a qualitative software analysis package [QSR International Pty. Ltd. QSR N5 (1991–2000)] to facilitate inductive content analysis (Miles and Huberman, 1994).

More in-depth and multiple (two to three) interviews were also conducted with a small sample of obese adolescents, which were also supported by parental interviews. Criterion sampling was used to recruit these obese young people and their parents from the obesity clinic. Patients were identified who (a) were female, (b) were adolescents (young people aged 12 to 18 years), and (c) had English as a first language, which was necessary for in-depth interviewing. The five participants [mean age: 15 (±2) years old] were information-rich participants as identified from previous involvement in a physical activity assessment project and familiarity with the interviewer. All interviews were audio-recorded and ranged in duration from 45 to 120 min for adolescents and from 30 to 90 min for parents. The gap between each participant interview was 1 wk. These interviews explored global and physical self-perceptions, peer relationships and victimization experiences, and weight management behaviors, specifically physical activity behavior. The mothers' interviews explored their children's psychological well being, peer relationships, and lifestyle behaviors. Interpretative Phenomenological Analysis (IPA) (Smith and Osborn, 2003) was used for analysis of the interview transcripts. Through interpretative, intensive engagement with texts and transcripts, IPA aims to explore meaningful experiences of individuals and is therefore concerned with trying to understand an individual's personal world through their perceptions or accounts of states or experiences (Willing, 2001).

FINDINGS

What Are the Physical Activity Experiences of These Obese Young People?

Half of the families reported that they had received some advice on how physical activity could help to reduce body weight. Advice had also been given, although limited, on how to increase their activity levels. Many obese young people described the inherent difficulties in being physically active when severely obese and how this promotes a negative cycle for weight management:

> My physical problems stop me from doing activity which means
> I put on more weight which means I can't do more activity, which
> means I put on more weight and it just goes on and on and on.

One key finding was the problem of dealing with physical size, aptly described by one participant that being physically large was "like, you know, a thin person going around with somebody piggybacking all the way along every day," and that they would feel happier if they could display greater levels of competence, particularly physical functioning, relative to their peers:

> ... you can feel more happy about doing sports when you're fit and you can last longer and you're not huffing and puffing all the time and you don't have to run at two miles an hour and everyone's like half up the field before you even get round the first corner and it's ... It's a lot better to be fit than not to be fit and be miserable ... it stops you from being normal. It's like being middle aged as a child. It's like, well yeah, it's like being old before your time. You can't ... you can't do everything that everybody else can do ...

Low physical condition, as reported by these obese young people, was particularly evident for aerobic activities of extended duration, where difficulties in "keeping up" with slimmer classmates were consistently reported by participants. Usually, the difference in size rather than the difference in conditioning was cited as the reason for not keeping up. Practical barriers related to location and access to facilities was also commonly reported (30% of participants). This was particularly the case for gyms, which often have age restrictions such that young adolescents are often not able to use the facilities. Even if gym access was not a problem, cost was cited as a prohibitive:

> A lot of doctors will say, oh go to the local gym ... But a lot of people can't afford it ...

Standard factors related to lifestyle for both parents and adolescents were also reported. Common factors related to time, the home-based approach to leisure, and the fragmented family life of some of these young people:

> Families don't seem to go out anymore and we've found, I mean in the summer holidays when I was a kid, in the summer holidays the streets were full of kids. You can't hear one these days. There's no one for her to play with because they are all either at holiday clubs or the parents have split up and they are with the father or the grandmother ....

Other psychological factors included previous unsuccessful weight loss attempts leading to subsequent abandonment of weight management strategies and weight gain, suggesting an "all or none approach" to weight management involvement. Unrealistic goal setting provoked low self-motivation and depleted enthusiasm.

Although the most frequently reported barriers tended to be the more obvious physical and practical barriers. Other barriers, although less frequently reported, appeared to have more profound impact on the individuals concerned. These barriers were generally described in relation to peoples' "attitude" and the associated embarrassment felt in being physically active in a nonsupportive environment. This was particularly the case for swimming, where students frequently reported the intrinsic enjoyment of swimming but a reluctance to visit public pools:

> You know ... gets embarrassed at the swimming pool. He wants to go, but when he gets there he don't want to go in. I think it's because the boys at school tormenting him ...

> She loves swimming ... she is a very good swimmer ... a water baby. But, she stopped swimming and she kept saying that she didn't feel very well ... and actually it was because she didn't want to get changed ... she was very conscious about getting undressed in front of other people and that made me feel really bad ...

Other studies have reported similar findings, with bullying and teasing reported to be major barriers for students to be physically activity during PE classes and before and after school and on sports teams (Bauer et al., 2004). This represents a significant challenge to promoting physical activity in obese young people, where the active environment may actually lead to victimization and attacks on their physical self, rather than promotion of the physical self. Interestingly, similar findings were evident from our data for eating a "healthy'" diet where an obviously "healthy" lunch box could be seen as a target for victimization in young obese people. However, it is important to emphasize that we found limited evidence overall of the physical activity context being a critical trigger for victimization experiences in young people. Furthermore, some of these morbidly obese young people strongly emphasized the benefits of being involved in physical activity, particularly the social aspects:

> It makes you feel you can do something. Makes you part of something ... makes you feel good and you feel like you can be better ...

In order to evaluate physical activity needs, interviews also explored desired physical activity components of weight treatment programs. A number of suggestions were made for physical activity as part of treatment programs, including exercise variety and choice, involvement with other obese peers, a gradual increase in the intensity level of the program,

Table 5.1   Key Inhibitors and Promoters for Physical Activity in Severely Obese Young People

Inhibitors of Physical Activity	Promoters of Physical Activity:
Peer exclusion ("non-supportive group environment")	Peer acceptance ("enjoyable group environment")
High-intensity activities leading to joint pain, breathing difficulties	Noncompetitive group-based activities (strength-based activities)
Competitive/comparative activities ("being last")	Non-"scoring" or individual rating activities
Not believing activity will make a difference	Clear goal setting and structure
Clothing requirements (being "on show")	Flexible approach to clothing
Unrealistic parental expectations; "told it's a good idea"	Parental support and participation
Limited motivation or "readiness to change"	
Physical condition and appearance	
Lack of time and facilities	

detailed individualized activity advice, support and encouragement, and regularity of these programs (i.e., weekly if possible). From these data, we have summarized in Table 5.1 key inhibitors and promoters for physical activity in this group of severely obese young people.

Although the potential effectiveness of physical activity was acknowledged, these obese young people and their families often attributed their physical status to emotional or external factors, genetic susceptibilities, or physiological factors beyond their control. This directly inhibited concerted efforts for healthy dietary and physical activity practices in many participants.

How is the Physical Self Related to Physical Activity Behavior?

Physical conditioning has already been mentioned as a significant barrier to physical activity participation in obese young people. Adolescent girls particularly emphasized the importance of this component of the physical self for functionality during day-to-day living and involvement in more structured forms of physical activity such as sports and exercise. All participants felt that they had low levels of physical conditioning and reported suffering tiredness and breathlessness. These issues and concerns impacted on everyday physical activity and, more specifically, within physical education classes, where comparisons were frequently made with fitness levels of peers. Comparisons of peer performance had the potential to provoke displeasure for involvement, specifically within high aerobic-high intensity and competitive activities. Low levels of physical condition generally inhibited, or reduced, participation in physical activity. This provoked frustration, and, as such, all participants reported their desire for improved physical conditioning.

All participants felt physically strong in comparison to their nonoverweight peers. A number of advantages of having physical strength and confidence in situations requiring strength were suggested. These included enhanced participation in strength-based physical activities, such as carrying shopping and power sports (track and field throwing); the provision of a sense of security and protection of self and friends; and the ability to defend themselves when confronted with overt bullying. However, perceptions of a large and muscular build resulted in feelings of nonfemininity.

Diversity in levels of perceived physical competence was established between participants. An obese status was not perceived to necessarily inhibit physical capability, and two of the adolescent girls enjoyed participating in sport, although participation in endurance (high-intensity) sports was reported to be more difficult. Other participants suffered greater physical difficulties, attributed to obesity comorbidities (breathing difficulties, back and joint pains) that directly impacted on their ability to be physically active.

Physical attractiveness in relation to size was important for the majority of participants. However, one isolated participant reported unique views of confidence, contentment, and self-acceptance of her size and appearance and higher overall self-worth, perhaps contributing to these positive self-evaluations. Acknowledged health complications, such as high blood pressure, were reported to be the primary reasons for weight loss desire rather than body dissatisfaction in this participant. In contrast, the other participants expressed high degrees of body dissatisfaction, in line with previous work (Neumark-Sztainer et al., 1999). Degrees of body dissatisfaction varied from self-consciousness and body-embarrassment to self-disgust. Social physique anxiety was an issue reported by all participants, especially within physical education, in which clothing, e.g., shorts or a swimming costume, that exposed their body was required. Clothes therefore played a functional role to cover the body. However, a prominent problem of being obese was the limited choice and availability of large clothing sizes.

Participants displayed heightened awareness of their self-presentation and physical body. Self-consciousness clearly affected self-confidence, which inhibited their ability to make acquaintances with peers and lead independent lives, away from the security of family members. However, although close friends provided a sense of security, negative self-evaluations influenced their feelings of worthlessness within these friendship dyads and resulted in feelings of social isolation. All participants reported the desire or necessity to lose weight. Perceived advantages of weight loss included the ability to fit in more and enhanced self-confidence and self-acceptance. "Embodied-largeness"(Carryer, 2001) was reflected in views that their obesity impinged on everything in their lives.

Surprisingly, given the centrality of weight concerns in the lives of the female participants, mothers suggested that their daughters spoke little about their feelings related to their weight. However, to varying degrees, all mothers referred to low self-confidence and high degrees of self-awareness in their daughters, especially during physical activity. Mothers were highly aware of their daughter's discontent with their obese status. They also acknowledged other physical self-perceptions, including physical limitations of living with an obese body. Mothers who had been obese themselves during childhood placed greater emphasis on these problems, perhaps reflecting their own "lived experience" of obesity.

CONCLUSIONS/IMPLICATIONS

Ways to Enhance Self-Perceptions Through Physical Activity

This chapter provides further support for the association between physical appearance and physical activity on global self-esteem (Sonstroem, 1997; Fox, 2000; Harter, 2001). However, the experience of physical activity, ranging from everyday activities, such as walking, to sport participation within school, clearly projected both positive and negative feelings for these obese participants. The following presents possible strategies to enhance physical education and other exercise experiences of obese adolescents and their self-perceptions of physical capabilities:

- Opportunities should be provided for strength-based activities. These were activities for which all participants held perceptions of competence, in comparison to high-intensity endurance activities, such as running.
- Exposing the body by wearing shorts in PE classes provoked anxiety and feelings of self-consciousness for four of the five participants. Physical activity leaders should provide more flexibility in clothing choice and allow, for example, jogging pants to be worn.
- Children should be encouraged to participate in small, relaxed groups with people they know and feel comfortable with.
- Mastery, self-determination, and ownership for success should be promoted.
- Opportunities should be provided to fulfill individual motives for physical activity involvement.
- Barriers to individual involvement should be explored by physical education coordinators and parents, and attempts should be made to reduce barriers.
- Aspirations should be modified: unrealistic expectations about athletic abilities can reduce global self-esteem. More appropriate goals,

including realistic and achievable competence levels, should be encouraged to modify unrealistic expectations about athletic abilities and how they impact on self-esteem and weight.

■ Physical activity leaders should be aware of risks of victimization for obese adolescents within the setting of physical education, which can provoke peer comparisons and highlight physical limitations.

■ Physical activity leaders should be aware of the physical problems associated with childhood and adolescent obesity, for example, orthopedic disorders of the hips and knees, which may limit physical activity involvement.

■ A nonthreatening and noncompetitive environment should be promoted, incorporating tolerance for error or lower levels of skill and ability.

Ways to Enhance Weight Management

Individual characteristics and behaviors play a key role in weight management; consequently, a doctrinaire global approach to treatment may not address these individual factors. As such, an individual approach to treatment based on personal characteristics and environmental factors, such as Cognitive Behavioral Therapy (CBT) (Cooper and Fairburn, 2001), may help to enhance weight loss and maintenance. The relationship between self-perceptions and behavior clearly suggests that evaluating personal self-perceptions is an essential preliminary stage for enhancing behaviors and individual motivating factors efficacious for weight management. This approach could also identify those young people who are ready to engage in weight management strategies and thus believe they hold self-control of their body weight. Obese young people need to be able to accept their ability to influence their weight status; however, if they internalize blame too much, then repeated failure may have an even more damaging impact on their sense of self. Setting realistic goals is an important component of weight management (Gibson et al., 2002). This is an important process to optimize the adoption of healthy dietary and physical activity practices, as recommended by the Expert Committee on Childhood Obesity (Barlow and Dietz, 1998). Although parental support and parenting skills provide a "foundation for successful intervention" (Barlow and Dietz, 1998), assessing motivations of the obese adolescents may reduce parent-child conflicts and secretive maladaptive practices, such as binge eating. Furthermore, maintenance of adaptive practices should be encouraged to enhance feelings of self-determination and achievement. In turn, peers also have a role to play in creating a supportive environment to reduce, for example, perceptions of physical incompetence within physical education and the subsequent desire to exclude themselves from involvement in weight management behaviors.

This focus on qualitative data in obese young people and their families highlights the complex interplay of factors that can inhibit or promote lifestyle behavior change. Understanding these cognitive, psychological, behavioral, and familial factors and being able to consistently manipulate them in clinical and school-based settings remains a real challenge for effective health promotion in obese young people.

ACKNOWLEDGMENT

We express our gratitude to all families who shared their experiences and also to the staff at the obesity clinic: Ruth Williams, Matthew Sabin, Julian Shield, and Elizabeth Crowne.

REFERENCES

Barlow S.E. and Dietz W.H., Obesity evaluation and treatment: Expert committee recommendations, *Pediatrics*, 102, e29, 1998.

Bauer, K.W., Yang, Y.W., and Austin., S.B., How can we stay healthy when you're throwing all of this in front of us? Findings from focus groups and interviews in middle schools on environmental influences on nutrition and physical activity, *Health Educ. Behav.*, 31, 34–36, 2004.

Biddle, S. and Armstrong, N., Children's physical activity: An exploratory study of psychological correlates, *Soc. Sci. Med.*, 34, 325–331, 1992.

Biddle, S. and Fox, N., Motivation for physical activity and weight management, *Int. J. Obes.*, 22, S39–S47, 1998.

Burrows A. and Cooper M., Possible risk factors in the development of eating disorders in overweight pre–adolescent girls, *Int. J. Obes.*, 26, 1268–1273, 2002.

Campbell, J.D. and Lavallee, L.F., Who am I? The role of self-concept confusion in understanding the behavior of people with low self-esteem, in *Self-Esteem: The Puzzle of Low Self-Regard*, Baumeister, R.F., Ed., Plenum Press, New York, 1993.

Carless, D. and Fox, K.R., The physical self, in *Interventions for Mental Health: An Evidence-based Approach for Physiotherapists and Occupational Therapists*, Everett, T., Donaghy, M., and Feaver, S., Eds., Butterworth-Heinemann, London, 2003.

Carryer, J., Embodied largeness: a significant women's health issue, *Nursing Inquiry*, 8, 90–97, 2001.

Cole, T.J., Bellizi, C., Flegal, K.M., et al., Establishing a standard definition for child overweight and obesity worldwide: international survey, *BMJ*, 320, 1–6, 2000.

Cooper, Z. and Fairburn, C.G., A new cognitive behavioral approach to the treatment of obesity. *Behav. Res. Ther.*, 39, 499–511, 2001.

Crocker, P.R.E., Eklund, R.C., and Kowalski, K.C., Children's physical activity and physical self-perceptions, *J. Sports Sci.*, 18, 383–394, 2000.

Delany, J.P., Harsha, D.W., and Kime, J.C., Energy expenditure in lean and obese prepubertal children, *Obes. Res.*, 3, 67–72, 1995.

Edmunds, L., Waters, E., and Elliott, E.J., Evidence based management of childhood obesity, *BMJ*, 323, 916–919, 2001.

Epstein, L.H., Myers, M.D., Raynor, H.A., et al., Treatment of pediatric obesity, *Pediatrics*, 101, 554–570, 1998.

Fox, K.R., The effects of exercise on self-perceptions and self-esteem, in *Physical Activity and Mental Well-Being,* Fox, K.R. and Biddle, S., Eds., Routledge, London, 2000.

Fox, K.R. and Edmunds, L., Understanding the world of the "fat kid": can schools help provide a better experience? *Reclaiming Child. Youth,* 9, 177–181, 2000.

French, S.A., Story, M., and Perry, C.L., Self-esteem and obesity in children and adolescents: a literature review, *Obes. Res.,* 3, 479–490, 1995.

French, S.A., Perry, C.L., Leon, G.R., et al., Self-esteem and change in body mass index over 3 years in a cohort of adolescents, *Obes. Res.,* 4, 27–33, 1996.

Gibson, P., Edmunds, L., Haslam, D.W., et al., *An Approach to Weight Management in Children and Adolescents (2–18 years) in Primary Care,* Royal College of Pediatrics and Child Health, National Obesity Forum, London, 2002.

Glenny, A.M. and O'Meara, S., *Systematic Review of Interventions in the Treatment and Prevention of Obesity,* York Publishing Services, York, 1997.

Harter, S., Is self-esteem only skin-deep? The inextricable link between physical appearance and self-esteem among American youth, *Reclaiming Child. Youth,* 9, 133–138, 2001.

Jelalian, E. and Mehlenbeck, R., Peer-enhanced weight management treatment for overweight adolescents: some preliminary findings, *J. Clin. Psychol. Med. Settings,* 9, 15–23, 2002.

Maffeis, C., Zaffanello, M., Pinelli, L., et al., Total energy expenditure and patterns of activity in 8 to 12-year-old obese and non-obese children, *J. Pediatr. Gastroenterol. Nutr.,* 23, 256–261, 1996.

Miles, M.B. and Huberman, A.M., *Qualitative Data Analysis,* 2nd ed., Sage Publication, Thousand Oaks, 1994.

Miller, C.T. and Downey, K.T., A meta-analysis of heavyweight and self-esteem, *Personality Soc. Psychol. Rev.,* 3, 68–84, 1999.

Molnar, D., Physical activity in relation to overweight and obesity in children and adolescents, *Eur. J. Pediatr.,* 159, S45–S55, 2000.

Neumark-Sztainer, D., Story, M., Faibisch, L., et al., Issues of self-image among overweight African-American and Caucasian adolescent girls: a qualitative study, *J. Nutr. Educ.,* 31, 311–320, 1999.

Phillips, R.G. and Hill, A.J., Fat, plain, but not friendless: self-esteem and peer acceptance of obese pre-adolescent girls, *Int. J. Obes.,* 22, 287–293, 1998.

Smith, J.A. and Osborn, M., Interpretative phenomenological analysis, in *Qualitative Psychology: A Practical Guide to Methods,* Smith, J.A., Ed., Sage, London, 2003.

Sonstroem, R.J., Physical activity and self-esteem, in *Physical Activity and Mental Health,* Morgan, W.P., Ed., Taylor and Francis, Washington, 1997.

Story, M., School-based approaches for preventing and treating obesity, *Int. J. Obes.,* 23, S43–S51, 1999.

Summerbell, C.D., Waters, E., Edmunds, L., et al., *Interventions for Treating Obesity in Children (Protocol for a Cochrane Review),* vol. 2, Update Software, Oxford, 2001.

Trost, S.G., Kerr, L.M., Ward, D.S., et al. Physical activity and determinants of physical activity in obese and non-obese children, *Int. J. Obes.,* 25, 822–829, 2001.

Walker, L.L.M., Gately, P.J., Bewick, B.M., et al. Children's weight loss camps: psychological benefit or jeopardy? *Int. J. Obes.,* 27, 748–754, 2003.

Willing, C., *Introducing Qualitative Research in Psychology: Adventures in Theory and Method,* Open University Press, Buckingham, 2001.

Wilson, P., O'Meara, S., Summerbell, C.D., et al., The prevention and treatment of childhood obesity. *Quality and Safety in Health Care,* 12, 65–74, 2003.

Young-Hyman, D., Schlundt, D.G., Herman-Wenderoth, L., et al., Obesity, appearance, and psychosocial adaptation in young African American children, *J. Pediatr. Psychol.,* 28, 463–472, 2003.

Zabinski, M.F., Saelens, B.E., Stein, R.I., et al., Overweight children's barriers to and support for physical activity, *Obes. Res.,* 11, 238–246, 2003.

Zwiauer, K.F.M., Prevention and treatment of overweight and obesity in children and adolescents, *Eur. J. Pediatr.,* 159, S56–S68, 2000.

6

WHAT'S IT LIKE BEING THE PARENT OF AN OVERWEIGHT CHILD?

Laurel Edmunds

CONTENTS

INTRODUCTION

The prevalence of obesity in children continues to increase globally (Ebbeling et al., 2002; Lobstein et al., 2004). The most recent figures in the U.K. suggest that 8.6% of 6-year olds and 15% of 15-year olds are now obese, as assessed by the 1990 body-mass index (BMI) cutoffs (CMO, 2002). Parents are very influential in their children's lives and are likely to play an important role in managing their children's weight

(Gibson et al., 2002; Summerbell et al., 2004), although there is some conjecture about the exact nature of that role (McLean et al., 2003).

There have been a number of focus group studies that have aimed to prevent obesity in children by educating and informing parents. These have been mainly with U.S. ethnic minority groups and have included exploring maternal perceptions of children's overweight (Borra et al., 2003; Crawford et al., 2004; Thompson and Story, 2003) and exploring the relationship between lifestyle behaviors, obesity, and health (Sherry et al., 2004; Thompson et al., 2003). However, these studies did not report parental experiences with health professionals, although Crawford et al. (2004) did conclude that a traditional approach to nutrition counseling with Latina mothers was unlikely to be effective, as maternal perceptions about child overweight and cultural values were operating as barriers.

One U.S. study, that of Jain et al. (2001), used focus groups to explore how lower-income, African-American mothers perceived overweight in their preschool children and why they did not seem to be concerned. The mothers' explanations for their children's developing weight and barriers to effective weight management were also investigated. Mothers did not accept the health professional's classification of overweight and distrusted the growth charts. This discrepancy was evident when mothers used the term "obesity" only with respect to grossly obese children and did not describe any child they had known as "obese." Parents thought inactivity resulted from obesity (rather than vice versa) and were concerned more by inactivity and teasing/low self-esteem than size. Mothers tended to use social functioning and activity to judge the healthiness of their child's weight and so emphasized the need to strengthen their child's self-esteem to counteract teasing and proposed this approach as an appropriate strategy to help their children rather than preventing obesity per se.

However, little is known about parents' experiences of having overweight children and their interactions with U.K. health professionals. Awareness of parental opinions and experiences is an important, but frequently over-looked, aspect of obesity prevention, particularly with young children. This chapter investigates the experiences of parents of overweight children in their younger years using in-depth interviews.

METHODS

Methodology

Childhood obesity is a particularly sensitive subject. The scientific research paradigm assumes one truth, and random samples, objectively sought, are tested so that generalizations can be made (Marshall, 1996). The method chosen for this study was face-to-face interviews because this allowed exploration of this very sensitive and complex subject, with the maximum flexibility for the discussion of novel topics. The most appropriate approach to analyzing these data is that used in the social sciences. The researcher is integral to the research, and their influence on the data collection process and interpretations of findings is acknowledged. My own overweight (BMI 29) resulted in comments such as "we don't mind talking to you because you're overweight" (South West weight loss group). The purposive sample targeted in this study was restricted to parents of school-age children who had weight concerns about their children. The aim was to identify a broad range of themes from their interviews, and their veracities were then checked in follow-up focus groups of participating parents.

Approval was granted by the Applied Qualitative Research Ethics Committee in Oxford (AQREC no. A00.020) and the South West Local Research Ethics Committee (study no. 2000/4/53s).

Parents with children of school age (4 to 16 years) were recruited from two areas of the U.K. Recruitment was through health professionals, posters in primary care settings, and advertising in local papers. Because of the sensitive nature of the topic, continued recruitment proved problematic; therefore, individuals were also recruited directly from weight-loss groups and further advertising. Awareness of the study was also raised by articles in the local papers in both areas and appeals on local radio (South West) and local television (Central area). Those interested in taking part received an information sheet and a reply slip via health professionals or at weight loss groups or were invited to telephone the author for more information when responding to advertisements.

Before interviewing began, respondents could discuss the research either at a weight-loss group meeting or on the telephone; discussions typically took approximately 30 min. In-depth, semi-structured interviews were then conducted by the author in the respondents' own homes, which were audiotaped with their written consent. The interview schedule was piloted and included topics such as the child's or children's weight history in conjunction with the body shapes (Stunkard et al., 1983) (see Figure 6.1). Parents were shown the shapes and asked to identify the relevant body shape for different ages as they recalled their child's life history. This part of the interview was frequently supported by family photographs. Subject matters discussed in interviews extended to other family members and experiences with health professionals.

Analysis

Descriptive data were recorded, including documentation of children's ages and the current and past shapes of the children and other family members. All of the interviews were transcribed verbatim, with the transcript checked against the interview for accuracy. These were analyzed

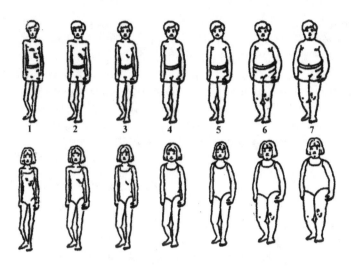

Figure 6.1   Body Shape Silhouettes of increasing obesity (after Stunkard et al., 1983.

using the method of constant comparison to identify emergent themes, which were coded for analysis. Themes were based on topics from the interviews and those emerging from the data (Pope et al., 2000; see Table 6.1). A sample of the transcripts was analyzed for themes by a second qualitative researcher, and resultant themes were defined and refined by the author during the periods of data collection and final analyses. Suitable quotes were selected to represent the range of views expressed and the recurrent themes rather than their frequencies, as these are potentially misleading within a purposive sample. Finally, two focus groups were held, one in each area, to test the veracity of findings; results were analyzed using the same method.

RESULTS

The sample consisted of parents of 40 children (20 from each area) with concerns about their child's weight. The numbers from each recruitment source are shown in Table 6.2. Their children were 4 to 15 years old, and 23 were girls. Parents came from a range of socioeconomic backgrounds and were classified using the Standard Occupational Classification (HMSO, 1998), shown in Table 6.3. Twenty-nine interviews were with mother alone, two with mothers and teenage daughters who wanted to be present, one where mother had a friend present, and the remainder with both parents. Three families were from ethnic minorities (Indian, Iranian, and Afro-Caribbean), and most children (*n* = 30) were living with both biological parents (four were with a nonbiological father, five were living with mother only,

Table 6.1   Themes Discussed from the Interview Schedule

Theme 1: Re-Establish the Rapport from the Telephone Conversation:
- Establish names, ages, birth order and children's schools
- Background questions about how each child is getting on at school, their hobbies and interests
- Establish a life grid for the family

Theme 2: Questions with Reference to the Life Grid:
- Weight history of each of the children covering their weight gain as a baby, their size when starting playgroup/nursery/school and at different points in school etc.
- The different body shapes were used to help parents to describe their child at different stages (four sets of body shapes were available: preadolescent, adolescent, adult male and adult female).
- Parents were also invited to have family photos available for the interview.
- For any period when the parent describes the child as overweight the interviewer will ask whether they were aware of the child being bothered about their weight and how the parents felt about it.

Theme 3: Questions Related to Parents:
- The parent's perception of the seriousness of weight gain in childhood, together with reference to their own weight history.
- Reasons for deciding to address, or not to address, their children's weight changes.

Theme 4: Social Interactions:
- Responses from others (such as friends, family, GP, practice nurse, teacher) were also sought on the child's weight.
- Experiences with health professionals were explored.
- If advice or help has been sought, or offered, from any source, how did the parent and child approach the issue, what went well and what went less well, and how could services be improved.

Theme 5: Possible Interventions:
- Parent's views on a range of innovative interventions for school children of different ages with weight problems were sought.
- Parental views of different approaches were sought.
- Parents were asked for ideas about how weight management should be addressed in children.

Theme 6: Bringing the Interview to a Close:
- Parents were asked if they have any other issues or questions they would like to raise, either on or off tape.
- Parents were asked if they were willing to take part in a focus group to discuss the findings of the study.

Table 6.2 Parental Recruitment Source by Area

Recruitment Source	South West	Central England
Pediatrician	0	0
GP (surgery)	0	1
Community Dietitian	0	0
Pediatric Dietitian	9(22 invited*)	0
Advertising	7	14
Slimming Groups	4	5
Total	20	20

* The pediatric dietitian setting provided the opportunity to invite parents to take part in the study by health professionals. This was not the case for other settings.

and one was adopted). The position of the child in the family also varied. Nine were only children, 12 were the eldest, 2 were middle children, and 16 were the youngest. Data for sex, age, and shape of the children are shown in Table 6.4. These variables have been used as identifiers (e.g., girl: 7y, shape 6) for parental observations.

Parents were aware that their babies and young children were larger than most and described them as "big," "chubby," "chunky," "plump," "fat," "tubby," and "podgy." Additionally, they made comments such as "he was always off the graphs," "always big," and "always had a weight problem from birth." Parents also had to cope with the attitudes of others and were subject to comments from strangers about their young children. In one

Table 6.3 Socio-Economic Backgrounds Based on the Occupation of the Main Parental Wage Earner in the Household

SES Class	South West	Central England
I	2	
II	7	15
III	1	4
IV	3	1
V	1	
Retired	4	
Student	1	
Unemployed	1	
Total	20	20

Table 6.4 Age, Sex and Shape of Children Estimated by Parents and the Author (Photographs were Available for All the Children)

Shape	4	5	6	7	8	9	10	11	12	13	14	15
3										•		
4												
4/5			•									
5		♦			•		♦	♦	♦	•	•	
5/6	•		•	•			♦	♦		•	♦	
6			•		•			♦♦		••		
6/7			•	♦		•♦	•	•	♦	♦	•	•
7				♦♦			♦	••	•		•	♦
Age	4	5	6	7	8	9	10	11	12	13	14	15

Girl = •; Boy = •

The normal shape, 13 year-old girl was included as her mother had future weight concerns for her daughter based on her own weight gain experiences.

instance, these experiences were so severe that a mother described that they "feel like lepers" (girl: 14y, shape 6/7). Some other comments were:

"Do they have the same father?"

[boy: 11y, shape 6 (with three preschool brothers)]

"Has she got a giantism problem?"

(girl: 13y, shape 5/6)

Negative societal attitudes where also apparent in school environments:

She looks like a big 6 year old. She's already aware that she is different, and a child at nursery told her she had a fat tummy. She's already sensitive about it.

(girl: 4y, shape 5/6)

He used to come home from school crying ... "please take me to the doctor Mummy, I'm too fat."

(boy: 5y, was shape 6)

Why don't you want to do PE, its fun? And she said "Somebody called me fat" ... and I think since then it's stuck in her head. You know, she's a bit self-conscious.

(girl: 6y, shape 6/7)

There were several apparent causes for children's overweight in their early years. These included: a high birth weight and children maintaining that size; inappropriate feeding practices in infancy; and medical intervention in infancy. One mother was convinced that her daughter's weight gain stemmed from misguided feeding advice:

> I was told at the postnatal clinic that she was losing weight and they were concerned ... I was trying to breast-feed her, but they said that if I didn't do something about her weight then she would have to go back into hospital. So they suggested SMA Gold. [Father] refers to this as "Miracle Grow" because as soon as she started drinking this stuff she just put on weight almost instantaneously and just put on weight and put on weight and put on weight. They [post-natal clinic] were very pleased with her progress and that she was putting on weight at the rate that she was. I think this may have been the beginning of her weight problem because she put on so much weight so quickly. Nobody reviewed the situation; it was just suggested that she carry on because she was doing so well on it.

(girl: 7y, shape 6)

Other parents questioned the impact of medical interventions when their children were young, particularly in relation to asthma treatments:

> He started getting a bit tubby when he started with his medicine [liquid] for asthma.

(boy: 8: shape 7)

> I did wonder whether it was the steroids; she'd come off the pills and ballooned. The doctor had said no, it was just normal puppy fat. I took her this time, and he was worried because she had gone so much bigger in such a short space of time.

(girl: 14y, shape 6/7)

Whatever the cause of the weight gain, mothers were left coping with their toddler's need for more food:

> She was always, always, always looking for food.

(girl: 9y shape 6)

> He used to go into the bin when he was little. A sort of desperation, a need for food.

(boy: 11y, shape 6)

She had hungry days. She would have her breakfast or lunch, and then two hours later she would be hungry again and you could hear her stomach rumbling.

(girl: 7y, shape 6)

Many of these parents had some help from health professionals for their child, but the reception they received was varied. Some doctors were positive and supportive even when they had little help to offer, and there was a tendency for health visitors and nurses to offer more practical advice. The pediatric dietitians with whom parents had contact were described as "lovely," "brilliant," and "very understanding." However, parents were frequently disappointed with the help they were offered:

My GP [general practitioner]—who's lovely and very helpful and I couldn't fault the way she handled it—but I think people are frightened to admit that, oh you might have a problem with an overweight child, people don't like talking about it.

(girl: 6y, shape 6/7)

He [registrar] was ever so sweet but he just didn't have a clue.

(girl: 9y, shape 6)

Pediatricians are dealing with kids all the time but they've got nothing ... they don't know what to do with overweight kids.

(boy: 8y, shape 7)

She [general practitioner] said "I don't know the answer."

(girl: 15y, shape 6/7)

Another approach was an attempt to allay parental concerns, which were not always successful:

"She's lovely; she will grow out of it ..."

(girl: 14y, shape 7)

"Oh stop being silly, he's fine, it's nothing." You leave there feeling like a paranoid parent ...

(boy: 10y, shape 5/6)

A contributory factor for the reluctance to address overweight in children was the concern about the unnecessary risk of precipitating an eating disorder. This was an issue for both health professionals and parents, but was frequently commented on by mothers:

> I worried that if I messed around with what she was eating, she might develop an eating disorder.
>
> *(girl: 9y, shape 6/7)*

> The doctor said it's more a case of tread very softly because you'll give her anorexia.
>
> *(girl: 12y, shape 7)*

Other health professionals conveyed more negative messages, resulting in mothers in particular feeling that they were "making a fuss" or were made to feel blameworthy or had their self-help approaches undermined:

> "These things all correct themselves" [general practitioner], but it's the traditional attitude of GPs that think you're being too fussy.
>
> *(boy: 11y, shape 6)*

> He [general practitioner] said, "You're not trying hard enough. You are doing something wrong. He should eat 700 calories a day."
>
> *(boy: 12y, shape 6/7)*

> GP to mother, "Well you buy the food."
>
> *(boy: 10y, shape 7)*

> I just couldn't believe she [dietitian] was the person that was nominated to give us advice because … she was so negative and it made us feel so inadequate …
>
> *(girl: 7y, shape 6)*

Many parents found that the help that they were offered left them feeling dissatisfied and the advice was little more than lifestyle homilies, which often showed a lack of understanding of their situation.

Everything that came home was bog standard, "try to do this and not that," we do already ...

(boy: 12y, shape 6/7)

... but there's no one to go to; like I said we've given up now because we don't get anywhere ...

(boy: 13y, shape 6/7)

I feel like I'm wandering around in the dark on my own ...

(girl: 14y, shape 6/7)

DISCUSSION

The views and experiences in this study are those of a purposive sample of parents who considered their child to have a weight problem; additionally, they were willing to discuss these experiences within a research framework. These views may or may not be representative of parents with overweight children generally. Similarly, their recall of what was actually said in consultations with health professionals may or may not be accurate. However, for these parents, the messages, impressions, and feelings that they associated with the consultations were their interpretations of reality, which were then impacting on their lives and those of their children. Their experiences were relatively consistent, both in how they coped with their children and at what point they sought professional help. The concern about medical implications began to emerge as the child approached shape 6, which is approximately equivalent to the 85th centile in the 1990 growth charts (Cole et al., 1995). In another interview-based study conducted with 100 overweight, but not obese, adults identified from medical records in primary care, participants raised medical-related weight concerns with children of this shape. Overweight in children was not the focus of the study, and participants considered none of their children to be overweight; however, ad hoc comments (56% of those interviewed) said shape 6 would be the size at which they would become concerned and seek professional help (Ziebland, 2001).

Childhood obesity is extremely complex, and several reasons for children's overweight were apparent in this study. Overweight and obese children tended to be regarded as a homogenous group, and so treatment strategies were mostly simple "eat less and do more messages," irrespective of the child's circumstances. The emphasis on dietary management may be due to the relative availability of dietitians and obesity being seen as a food problem by health professionals rather than the most appropriate strategy. Managing food intake in a child who is always hungry is likely to be a fraught experience and may create more problems for child and parent if no support is available beyond limited dietary counseling. Both health professionals and parents were concerned about precipitating eating disorders (Rainey et al., 1998), but children were still referred to dietitians. Although the evidence is equivocal (Killen et al., 1994; Patton et al., 1990; Schleimer, 1983), the emphasis on diet appeared to be a further deterrent for parents. Ironically, a consistent finding from the evidence suggests that a reduction of sedentary behaviors and increasing physical activity may be more effective (Campbell et al., 2003; Summerbell et al., 2004).

The responses from health professionals ranged from being sympathetic to being dismissive and were likely to be reflecting two factors: the social stigma attached to being overweight as a personal responsibility and the lack of a solid evidence base for treating childhood obesity in primary and secondary care. Despite the increasing prevalence of overweight and obesity in the population, the associated stigma appears to be worsening (Latner and Stunkard, 2003). If health professionals believe that parents are responsible for their children's body status without acknowledging all the environmental changes that have taken place that are driving a global increase in obesity, it is not surprising for all of their reactions to be unsympathetic. The responses from some of the health professionals reported by parents may reflect the findings of Jelalian et al. (2003), in which health professionals were found to feel a lack of competence with respect to childhood obesity.

Parents received little help beyond advice (that their child should eat healthily and do more exercise) and, in some cases, "doing tests," which were all negative in these children. This proved to be a source of frustration and/or resignation for both parents and health professionals alike. The advice currently available to health professionals is nonspecific and cautionary (Gibson et al., 2002; SIGN, 2003, Summerbell et al., 2004). However, with nearly one third of U.K. children either overweight or obese (CMO, 2002), health professionals are likely to be consulted in increasing numbers. In general, improved practitioners' attitudes and understanding of childhood obesity are necessary for future effective weight management advice. There are other approaches (e.g., psychological support, parenting support, counseling) that may be more helpful for some children, e.g., for the child that is grieving or for parents with pressing social problems (Edmunds et al., 2001; Gibson et al., 2002).

In summary, the interviewed parents were trying to balance their children's dietary and physical activity needs with those of family and without causing any further stigmatization of their overweight child. Many had to cope with health professionals who were less than sympathetic (with some who may have inadvertently contributed to the child's weight problem), and the level of help available to them was minimal. One of the most difficult aspects of childhood obesity is the associated social sensitivity and the reluctance of anyone to discuss it. The variety of responses that parents encountered from health professionals may be due to issues like not wanting to cause further distress to either parent or child or just not knowing how to help. Health care professionals also tend to perceive the parent as blameworthy and overweight as an individual rather than a societal problem. Further research is needed with health professionals to clarify their understanding of all the issues involved, but there would appear to be a need for guidance and training with respect to childhood obesity.

ACKNOWLEDGMENTS

This work was funded by South East Region NHS Executive Research and Development Fund (Grant SEO 151) and approved by the Applied Qualitative Research Ethics Committee Study A00.020 and the South West Local Research Ethics Committee Study No. 2000/4/53. We thank the parents who took part and the health professionals that helped, and Rosemary Conley Slimming Clubs and Slimming World for their support in my efforts to recruit respondents.

REFERENCES

Borra, S.T., Kelly, L., Shirreffs, M.B., et al., Developing health messages: qualitative studies with children, parents, and teachers help identify communications opportunities for healthful lifestyles and the prevention of obesity, *J. Am. Diet. Assoc.,* 103, 721–728, 2003.

Campbell, K., Waters, E., O'Meara, S., and Summerbell, C.D., Interventions for preventing obesity in children (Cochrane Review), in *The Cochrane Library,* John Wiley & Sons, Ltd., Chichester, 2003, issue 4.

CMO, *Annual Report of the Chief Medical Officer,* 2002, available at http://www.doh.gov.uk/cmo/annualreport2002/contents.htm.

Cole, T.J., Freeman, J.V., and Preece, M.A., Body mass index reference curves for the UK, 1990, *Arch. Dis. Child.,* 73, 25–29, 1995.

Crawford, P.B., Gosliner, W., Anderson, C., et al., Counseling Latina mothers of preschool children about weight issues: suggestions for a new framework, *J. Am. Diet. Assoc.,* 104, 387–394, 2004.

Ebbeling, C., Pawlak, D. and Ludwig, D., Childhood obesity: public-health crisis, common sense cure, *Lancet,* 360, 473–482, 2002.

Edmunds, L., Waters, E. and Elliot, E.J., Evidence-based management of childhood obesity, *BMJ,* 323, 916–919, 2001.

Gibson, P., Edmunds, L., Haslam, D.W., and Poskitt, E., An approach to weight management in children and adolescents (2–18 years) in primary care, *J. Fam. Health Care,* 12, 108–109, 2002.

HMSO Office of Population Census and Surveys, *Standard Occupational Classification,* HMSO, London, 1998, vol. 3.

Jain, A., Sherman, S.N., Chamberlin, D.L., et al., Why don't low-income mothers worry about their preschoolers being overweight?, *Pediatrics,* 107, 1138–1146, 2001.

Jelalian, E., Boergers. J., Alday, C.S., and Frank, R., Survey of physician attitudes and practices related to pediatric obesity, *Clin. Pediatr.,* 42, 235–245, 2003.

Killen, J.D., Taylor, C.B., Hayward, C., et al., Pursuit of thinness and onset of eating disorder symptoms in a community sample of adolescent girls: a three-year prospective analysis, *Int. J. Eating Disord.,* 163, 227–238, 1994.

Latner, J.D. and Stunkard, A.J., Getting worse: the stigmatisation of obese children, *Obes. Res.,* 11, 452–456, 2003.

Lobstein, T., Bauer, L., and Uauy, R., Obesity in children and young people: a crisis in public health. *Obesity Reviews* 5 (Suppl. 1), 1, 2004.

Marshall M., Sampling for qualitative research, *Family Prac.,* 13, 522–525, 1996.

McLean, N., Griffin, S., Toney, K., and Hardeman, W., Family involvement in weight control, weight maintenance and weight-loss interventions: a systematic review of randomised trials, *Int. J. Obes. Relat. Metab. Disord.,* 27, 987–1005, 2003.

Patton, G.C., Johnson-Sabine, E., Wood, K., et al., Abnormal eating attitudes in London schoolgirls-a prospective epidemiological study: outcome at twelve month follow-up, *Psychol. Med.,* 20, 383–394, 1990.

Pope, C., Ziebland, S., and Mays, N., Qualitative research in health care: analysing qualitative data, *BMJ,* 320, 114–116, 2000.

Rainey, C.J., Kemper, K.A., Poling, R., et al., Parents' perceptions of influences on child eating behaviors: an attitudinal approach, *J. Health Educ.,* 29, 223, 1998.

Schleimer K., Dieting in teenage schoolgirls. A longitudinal prospective study, *Acta Paediatr. Scand. Suppl.,* 312, 1–54, 1983.

Sherry, B., McDivitt, J., Birch, L.L., et al., Attitudes, practices, and concerns about child feeding and child weight status among socioeconomically diverse white, Hispanic, and African-American mothers, *J. Am. Diet. Assn,* 104, 215–221, 2004.

SIGN, *Management of obesity in children and young people,* Scottish Guideline Agency, 2003, available at http://www.Sign.ac.uk/guidelines/fulltext/69/index.html.

Stunkard, A.J., Sorensen, T.I., and Schulsinger, F., Use of the Danish adoption register for the study of obesity and thinness, in *Genetics of Neurological and Psychiatric Disorders,* Kely, S.S., Rowland, L.P., Sidman, R.L. and Matthysse, S.W., Eds,. Raven Press, New York, 1983, p. 115.

Summerbell, C.D., Waters, E., Edmunds, L., et al., Interventions for treating obesity in children (Cochrane Review), in *The Cochrane Library,* John Wiley & Sons, Ltd., Chichester, 2004, issue 1.

Thompson, V.J., Baranowski, T., Cullen, K.W., et al., Influences on diet and physical activity among middle-class African American 8- to 10-year-old girls at risk of becoming obese, *J. Nutr. Educ. Behav.,* 35, 115–123, 2003.

Thompson, L.S. and Story, M., Perceptions of overweight and obesity in their community: findings from focus groups with urban, African-American caretakers of preschool children, *J. Natl. Black Nurses Assn.,* 14, 28–37, 2003.

Ziebland, S., Unpublished data, 2001.

PART II

THE ROLE OF BIOLOGICAL AND SOCIAL PROCESSES IN THE AETIOLOGY OF CHILDHOOD OBESITY

7

THE VALUE OF BIRTH COHORTS IN THE STUDY OF CHILDHOOD OBESITY

Andy Ness

CONTENTS

INTRODUCTION

Children are becoming fatter, and the prevalence of obesity is increasing in the U.K. (Reilly et al., 1999; Reilly and Dorosty, 1999; Bundred et al., 2001; Rudolf et al., 2001a, 2001b; Lobstein et al., 2003) as it is across Europe (Livingstone, 2001), in Australia (Magarey et al., 2001), and in the U.S. (Ogden et al., 1997). This is despite secular increases in height. The scale of this increase is illustrated in Table 7.1 with data from the Avon Longitudinal Study of Parents and Children (ALSPAC), a large birth cohort of contemporary U.K. children (Reilly et al., 1999; Golding et al., 2001).

This recent increase in childhood adiposity has important immediate and long-term health implications (Must and Strauss, 1999; Must, 1996; Visscher

Table 7.1 Prevalence of Overweight and Obesity in the Avon Longitudinal Study of Parents and Children (ALSPAC) at Ages 24, 49, and 61 Months

	Age (months)		
	24	49	61
Total number	1031	1013	972
Number (%) overweight	163 (15.8)[a]	206 (20.3)[a]	182 (18.7)[b]
Number (%) obese	62 (6.0)	77 (7.6) [b]	70 (7.2) [b]

Note: ALSPAC recruited 14,541 pregnant women with an estimated date of delivery between 1st April 1991 and December 1992 and resident in one of the three Bristol-based health districts of Avon in the South West of England. These analyses are based on the follow-up of a 10% sample of children measured regularly in research clinics (called the children in focus) randomly selected from those born in the last 6 months of recruitment.

Note: Overweight and obesity were defined on the basis of body mass index [BMI, (weight in kg)/(height in m^2)] and compared with 1990 U.K. reference values. Overweight was defined as BMI above the 85th centile, and obesity was defined as BMI above the 95th centile. Thus, the expected prevalence of overweight was 15% and the expected prevalence of obesity was 5%.

[a] $p < .05$ [b] $p < .01$.

Based on data from: Reilly J.J., et al., BMJ, 319, 1039, 1999.

and Seidell, 2001). Although some markedly obese children suffer serious physical morbidity (in this respect, the recent reports of Type 2 diabetes in adolescents are of particular concern) (Fagot-Campagna et al., 2000; Drake et al., 2002), the most important immediate consequences may be psychosocial (Must and Strauss, 1999; Must, 1996). For example, ALSPAC data were recently used to examine the association between obesity and bullying behavior (Foster et al., 2004). These results are summarized in Table 7.2.

Evidence suggests that more obese children have a higher risk of asthma (Figueroa-Munoz et al., 2001; von Mutius et al., 2001). This phenomenon is independent of atopy (Figueroa-Munoz et al., 2001) and appears to have emerged only recently (Chinn and Rona, 2001), suggesting an association of airway reactivity with lifestyle characteristics underlying the obesity epidemic. Some adult disease processes begin in childhood (McGill and McMahon, 2003). For example, the Bogalusa Heart Study reported that fatty streaks and even atheromatous plaques are present in young children (Berenson et al., 1998). The Bogalusa and Muscatine studies have shown that cardiovascular risk factor levels in childhood predict the subsequent amount of atheroma found postmortem or radiographically visible coronary calcification (Berenson et al., 1998; Newman et al., 1986). Childhood obesity

Table 7.2 Adjusted Odds Ratio for Overt Bullying by Weight Categories in Boys and Girls in the Avon Longitudinal Study of Parents and Children (ALSPAC)

Boys	Number	Bully Odds Ratio	95%CI	Victim Odds Ratio	95%CI
Underweight ‡	391	0.90	0.56, 1.38	0.77	0.58, 1.03
Average weight	2273	1.00	—	1.00	—
Overweight	273	1.52	0.99, 2.36	0.90	0.64, 1.26
Obese	255	1.66	1.04, 2.66	1.54	1.12, 2.13
Girls	Number	Bully Odds Ratio	95%CI	Victim Odds Ratio	95%CI
Underweight[a]	290	1.58	0.86, 2.90	1.05	0.79, 1.39
Average weight	1808	1.00	—	1.00	—
Overweight	249	0.72	0.31, 1.71	0.84	0.61, 1.16
Obese	168	1.53	0.68, 3.44	1.53	1.09, 2.15

Note: Odds ratio was adjusted for parental (maternal and paternal) social class assessed at 32 weeks gestation before birth of child.

Note: Bullying was assessed at age 8.5 years using a structured face-to-face interview — the Bullying and Friendship Interview. Children were asked whether they had experience of overt bullying such as having belongings stolen, having been threatened or blackmailed, having been hit or beaten up, having been called bad or nasty names, and having nasty tricks played on them. The responses to these questions were used to classify children as overt bullies (children who were involved in overtly bullying of others frequently or every week), overt victims (children who experienced any form of overt bullying frequently or every week), and overt neutrals (children who neither physically bullied others nor became physical victims).

[a]At age 7.5+ years, height was measured to the nearest 0.1 cm using a Leicester Height meter and weight was measured to the nearest 0.1 kg using Seca model 835 scales. Underweight, overweight, and obesity were defined on the basis of BMI (weight in kg/(height in m²) and compared with 1990 U.K. reference values. Underweight was defined was defined as BMI below the 15th centile, overweight was defined as BMI above the 85th centile, and obesity was defined as a BMI above the 95th centile.

Source: Foster, L.J., et al., In preparation.

has in turn been positively associated with cardiovascular disease risk factor levels (Williams et al., 1992; Freedman et al., 1999) and more recently with carotid intima-medial thickness in adulthood (Li et al., 2003; Raitakari et al., 2003). Furthermore, some studies have reported a positive association between adiposity in childhood and subsequent adult cardiovascular and all-cause mortality (Gunnell et al., 1998; Must et al., 1992; Nieto et al., 1992). In addition to the effects on health, obese children (particularly older obese children) are at increased risk of becoming obese adults, although in the past many obese adults were not obese as children (Guo and Chumlea, 1999; Whitaker et al., 1997; Power et al., 1997). A recent study comparing body-mass index (BMI) in childhood with bioimpedance in adulthood challenged this view and suggested that tracking of BMI was due to tracking of lean mass (Wright et al., 2001). These results need to be confirmed in studies with serial measures of fat mass and lean mass, both of which may contribute to the tracking of BMI. Obesity in adulthood is associated with increased risk of coronary heart disease, stroke, and non-insulin-dependant diabetes mellitus (National Task Force on the Prevention and Treatment of Obesity, 2000).

Obesity is a condition in which fat stores are enlarged to an extent that impairs health (Garrow, 2000). The gold standard for assessment of body composition is cadaveric dissection. All other methods are indirect and require assumptions to be made (Ellis, 2000; Dewit et al., 2000; Wells et al., 1999). BMI is widely used to assess population levels of adiposity because it is easy to measure. Comparisons with bioimpedance measures in ALSPAC suggest that BMI is specific and moderately sensitive in identifying obese children, provided that appropriate cutoffs are chosen (Reilly et al., 2000). However, across the normal range, BMI does not distinguish well between fat and lean mass (Wells, 2000). This may be important because the determinants and consequences of each may be different in children, as suggested by data relating fat and lean mass to mortality in adults (Heitmann et al., 2000). Dual-energy x-ray absorptiometry (DXA) provides more accurate estimates of fat mass than either BMI or bioimpedance and also estimates lean mass and bone mass (Davies, 1997; Wells et al., 1999). DXA also provides information on the regional distribution of body fat, which may be particularly important (Després et al., 2001). It will be interesting to see the results of measuring body composition with whole body DXA at ages 9.5+ and 11.5+ in ALSPAC as well as at ages 13+ and 15+.

Obesity is a result of chronic energy imbalance (Weinsier et al., 1998; Wells, 1998). Homeostatic mechanisms coupling diet and physical activity appear to be better for energy deficits than for energy excess (Moore, 2000). Studies of the family history of obesity as well as adoption and twin studies of obesity demonstrate that individual susceptibility has a strong genetic component (Stunkard et al., 1986; Sorensen et al., 1992; Parsons et al., 1999; Sorensen and Echwald, 2001). Genetic make up cannot, however, explain the recent increases in adiposity in children.

Diet survey data seem to show that population levels of obesity have increased in the face of declining energy intake, implying that inactivity may be more important than overconsumption (Weinsier et al., 1998; Prentice and Jebb, 1995). In contrast, food availability data, which may be less susceptible to bias than individual level data, suggest that energy intake has not declined and may have even increased in the U.S. (French et al., 2001).

Studies of obesity cannot measure usual energy intake or usual energy expenditure in free-living humans and thus cannot identify whether this imbalance is the result of increased consumption or reduced expenditure. Free-living human obesity studies are thus seeking to describe dietary factors and patterns of physical activity that predispose to or protect against energy imbalance. Some individual based studies have reported positive associations between obesity and sedentary activities, such as watching television, in children (Dietz and Gortmaker, 1985; Andersen et al., 1998; Dietz and Gortmaker, 1993; Robinson, 1999), whereas others have reported associations with certain aspects of diet, such as saturated fat intake or the consumption of sweetened drinks (Ludwig et al., 2001). A systematic review carried out by Parsons et al. concluded that the protective effects of physical activity in childhood against later obesity was "inconsistent but suggestive" and that "studies investigating diet in childhood were limited and inconclusive" (Parsons et al., 1999). Thus, there is a need for large, longitudinal studies with accurate measures of physical activity, diet, and body composition in childhood.

A recent joint WHO/FAO report on diet, nutrition, and the prevention of chronic diseases reviewed the evidence (Report of a Joint WHO/FAO Expert Consultation, 2003). This report concluded that there was convincing evidence that regular physical activity and high intake of nonstarch polysaccharides (dietary fiber) reduce the risk of obesity and that sedentary lifestyles and high intake of energy-dense micronutrient-poor foods increase the risk of obesity. The report also concluded that there was evidence that home and school environments that support healthy food choices for children and that breastfeeding probably reduce the risk of obesity; heavy marketing of energy-dense food and fast-food outlets, high intake of sugar-sweetened soft drinks and fruit juices, and adverse socioeconomic conditions (in developed countries, especially for women) were also shown to probably increase the risk of obesity.

EPIDEMIOLOGICAL APPROACHES TO THE STUDY OF OBESITY

Evidence from epidemiological studies is complementary to that from detailed clinical studies, qualitative studies, and laboratory experiments. Together, these sources of evidence should help us to understand the causes of childhood obesity and identify potential interventions to prevent or treat childhood obesity.

Ecological studies compare the characteristics of groups rather than of individuals. One of the advantages of such studies is that the differences in exposures between populations may be far more marked than between individuals. A major disadvantage of such studies is that the associations observed may be due to confounding as different populations differ in many ways from each other. In a recent study of U.K. children that compared data measured between 1989 and 1999 with reference data from surveys performed between 1963 and 1975, BMI was similar; however, children in the 1990s had a greater fat mass and a lower lean mass (Wells et al., 2002). The findings of this study require confirmation but suggest that the balance between fat and lean may vary across populations and over time. Despite these potential difficulties, an examination of the power of associations from prospective studies to explain time trends and geographical differences is a valuable check on the importance of these associations.

Case series report on observations within a number of cases and can provide valuable information on the emergence of new conditions or on the changing nature of disease. An example of a case series is the recent report of four cases of Type 2 diabetes in obese white children aged 13 to 15 years (Drake et al., 2002). Case series lack a clearly defined reference population, which can be particularly problematic in cases drawn from a tertiary referral center, and do not include a control group for comparison. Furthermore, it may not be clear with diseases such as Type 2 diabetes (that are often initially asymptomatic) whether there has been an increase in the incidence of disease or whether cases are now being recognized and recorded or better treated or both (Gale, 2003).

Case control studies compare individuals with disease with individuals without disease. Case control studies such as those on activity patterns in normal, overweight, and obese adults (Cooper et al., 2000) and in normal, overweight, and obese children (this volume) can provide potentially interesting insights into differences, for example, between children with obesity and children who are not obese. There is, however, always the possibility that associations are explained by bias in selection of cases and controls or in the information collected on exposure. Furthermore, it is possible that children with obesity have modified their activity levels as a result of becoming obese and that this explains the observed association.

Randomized, controlled trials represent the gold standard for establishing causal relationships in medical research, but they are not always possible (Black, 1996). Observational studies can help select the interventions that are most likely to result in a benefit and to identify situations where trials may not be feasible.

Prospective studies have a number of advantages over other observational studies of childhood obesity. Unlike clinical case series where the population at risk is not well defined, the cases in a cohort study arise from a clearly identified group. And because exposure is measured before obesity develops, the potential for bias is reduced. As with all observational studies, confounding is an issue. This is a particular problem for socially patterned outcomes. Because a number of behaviors are socially patterned, it makes it difficult

Table 7.3   Unadjusted Odds Ratio for Obesity at Age 7.5+ by Level of Maternal Education in the Avon Longitudinal Study of Parents and Children (ALSPAC)

Maternal Education[a]	Odds Ratio	95% CI
CSE or less	2.57	1.83, 3.61
Vocational	2.65	1.84, 3.83
O level	1.70	1.24, 2.31
A level	1.74	1.26, 2.39
Degree	1.00	—
P for trend		<0.0001

Note: For definition of obesity see footnote to Table 7.2

[a] Self-reported maternal education at 32 weeks gestation before birth of child

Source: Reilly J.J., Armstrong, J., Sherriff, A., Dorosty, A., Emmett, P., Ness, A.R., and the ALSPAC study team, unpublished observations.

to tease out which if any of the behavior-obesity associations is causal. Obesity is socially patterned in developed countries (Report of a Joint WHO/FAO Expert Consultation, 2003). The social patterning of childhood obesity is illustrated in Table 7.3 using data from ALSPAC.

BIRTH COHORTS

Birth cohorts are prospective studies that recruit children born over a defined period, usually at or before birth, and follow these children up over time. Several large national or regionally based birth cohorts have been set up in the U.K. in the past 50 years, and large birth cohorts have been set up or are planned across the world. Details of some of these cohorts are summarized in Table 7.4.

Birth cohorts can look at the association of childhood obesity with cumulative exposures. They can also examine the observed associations with exposures at critical time periods.

From epidemiological studies, Dietz (1994) suggested that there were three critical time periods for the development of childhood obesity: the prenatal period (Whitaker and Dietz, 1998), the period in early childhood when adiposity rebound occurs (Rolland-Cachera et al., 1994), and adolescence (Dietz, 1994). Biological studies have further suggested that the growth curve of adipose tissue is maximal at birth and just before the onset of puberty

Table 7.4   Examples of Birth Cohorts (in Chronological Order of Recruitment)

Name of Study	Population Recruited	Number Recruited
MRC National Survey of Health and Development[a]	Class stratified sample of children born in one week in 1946 (3rd to 9th March)	5,362 children
National Child Development Survey[b]	Children born in England, Wales and Scotland in one week in 1958 (3rd to 9th March)	17,773 children
British Cohort Study 1970 (BCS70)[b]	Children in the United Kingdom born in one week in 1970 (5th to 11th April)	16,135 children
Birth to Twenty Study (formerly the Birth to Ten Study)[c]	Births during a 7-week period in Soweto and Johannesburg (April to June 1990) to resident mothers	3,273 children
The Avon Longitudinal Study of Parents and Children (ALSPAC)[d]	Pregnant women with an estimated date of delivery between 1st April 1991 and December 1992 and resident in one of the three Bristol-based health districts of Avon in the South West of England	14,541 pregnant women and 14,676 fetuses
Project Viva[e]	Women with singleton pregnancies attending one of eight offices of Harvard Medical Associates (a large multi-specialty group practice) in Eastern Massachusetts between April 22nd 1999 and July 31st 2002	2,341 pregnant women and 2,128 live infants
The Danish National Birth Cohort (DNBC)[f]	Pregnant women in Denmark from 1996–2002	101,046 pregnant women
Norwegian Mother and Child Study[g]	Pregnant women in Norway	33,000 pregnant women to date; aim to recruit 100,000 by the end of 2006

Source: [a]Wadsworth, M. and Hardy R., in *National Heart Forum, A Lifecourse Approach to Coronary Heart Disease Prevention. Scientific and policy review,* Stationery Office, London, 2003; [b]Ferri E., *Paediatr. Perinat. Epidemiol.* 12S, 31–44, 1998; [c]Richter, L.M., et al., *Paediatr. Perinat. Epidemiol.,* 18, 290–301, 2004; [d]Golding, J., et al., *Paediatr. Perinat. Epidemiol.* 15, 74–87, 2001; [e]Gillman, M.W., et al., *J. Paediatr.,* 144, 240–245, 2004; [f]Olsen, J., et al., *Scand. J. Public Health,* 29, 300–307, 2001; [g]Magnus P., Protocol for the Norwegian Mother and Child Study, available at www.fhi.no/tema/morogbarn.

(Wabitsch, 2000). Dietz and Gortmaker (2001) are now less convinced that birth weight per se is important and concluded recently that the "relevance of each of these periods to the prevalence of adult obesity remains uncertain."

Higher birth weight (a composite measure of both fat and lean mass at birth) is associated with increased BMI in later life, but there is also some evidence that low birth weight is also associated with increased obesity (particularly central adiposity) (Oken and Gillman, 2003). In the Dutch hunger winter, the offspring of women exposed to famine during the first trimester of pregnancy were more likely to become obese, whereas those exposed to famine in the third trimester of pregnancy were less likely to become obese (Ravelli et al., 1976). The recent finding that birth weight was more strongly associated with lean mass than fat mass in a study of 143 older people (Gale et al., 2001) may explain the association between birth weight and BMI reported previously (Whitaker and Dietz, 1998). Several studies have reported that children born to smokers were subsequently fatter (Vik et al., 1996; Power and Jefferis, 2002). If this association is causal this would show that an environmental insult in pregnancy can indeed affect subsequent fat mass.

After an increase in the first year of life, BMI falls until the age of about 4 to 6 years after which it increases again. This increase in BMI is called adiposity rebound, and an early rebound has been shown to be associated with higher BMI in adults (Dietz, 1994; Rolland-Cachera et al., 1994; Dietz, 2000). Adiposity rebound has not, however, been shown to be associated with a change in fat mass or later differences in fat mass. (Dietz, 2000) Patterns of growth in early childhood may nevertheless determine subsequent risk of becoming obese (Ong et al., 2000; Parsons et al., 2001). In an analysis of the ALSPAC cohort, rapid growth in the first 2 years of life was found to be associated with obesity at age 5 years (based on BMI and skin-fold thickness) (Ong et al., 2000). Others have reported that high growth rates in the first 7 years of life were associated with increased BMI at age 33 years (Parsons et al., 2001).

The amount and proportion of fat and lean mass change through puberty. These changes are different for boys and girls (Maynard et al., 2001). In boys, fat mass declines from age 14 years and then rises again at about age 17 years, whereas lean mass increases from about 13 to 14 years of age onward (Maynard et al., 2001). In girls, fat mass increases from age 8 to 18 years, whereas the increase in lean mass is much less marked than in boys (Maynard et al., 2001). Maturational timing has been consistently shown to be associated with obesity—with those having early onset of puberty becoming more obese (although there is more data for girls than boys because of ease of measuring onset of periods in girls as a marker of pubertal timing) (Garn et al., 1986; van Lenthe et al., 1996; Okasha et al., 2001;

Table 7.5 Breastfeeding and Risk of Obesity at Age 7.5+ Years in the Avon Longitudinal Study of Parents and Children (ALSPAC)

Breastfeeding	Unadjusted Odds Ratio	95% CI	Fully Adjusted Odds Ratio	95% CI
Exclusive at 2 months	0.64	0.50, 0.82	1.22	0.87, 1.71
Stopped/nonexclusive	0.81	0.64, 1.01	1.08	0.80, 1.45
Never breastfed	1.00	—	1.00	—
P for trend		0.002		0.464

Note: For definition of obesity see footnote to Table 7.2. Breastfeeding history was based on a self-completed questionnaire sent to mothers around 6 months postpartum. Odds ratio was adjusted for gender, energy intake at 38 months, and maternal education (for categories, see Table 7.3), as well as birth weight (continuous), maternal smoking at 28–32 weeks gestation (smoking categories: none, 1–9/day, 10–19/day, 20+/day), timing of introduction of solids (categories variable: <1 month, 1–2 months, 3–4 months, 4–6 months), parental obesity (categories: both parents BMI <30, father BMI 30, mother BMI 30, both parents ≥ 30), number of siblings (categories: none, one, two, three or more), hours watching television per week on questionnaire at 38 months (categories: 4 hours, 4–8 hours, >8 hours), hours of sleep overnight at 30 months (categories: <10.5 hours, 10.5–10.9 hours, 11–11.9 hours, >12 hours), and childhood dietary patterns (four variables for different dietary patterns derived by principal components analysis and termed "junk," "healthy," "traditional," and "fussy/snack" and analyzed as quartiles of each score).

Source: Reilly, J.J., et al., unpublished observations.

Must et al., 1999). In addition, the NHLBI growth and health study showed that the differences in obesity between the black and white U.S. girls emerged during puberty and that the early onset of menarche in the black girls was partly explained by their higher fat mass (Kimm et al., 2001). Whether these associations with pubertal onset represent an effect of maturational timing on fat and lean mass or are an effect of fat and lean mass on maturational timing requires investigation in longitudinal studies with accurate measures of body composition before and after puberty.

Several studies have suggested that breastfeeding may protect against subsequent obesity (Armstrong and Reilly, 2002; Ebbeling et al., 2002; Parsons et al., 2003). Birth cohorts may have collected more detailed information on possible confounders and so be better able to adjust for confounding. In both the 1958 cohort (Parsons et al., 2003) and ALSPAC, there was a protective association between breastfeeding and obesity that was no longer apparent after adjustment. The unadjusted and adjusted associations between breastfeeding and obesity in ALSPAC are shown in Table 7.5.

THE CHALLENGES OF BIRTH COHORTS

The use of birth cohorts has a number of challenges. These are summarized in Table 7.6. The challenge of expense and piecemeal project funding is probably best solved by setting up different mechanisms to ensure adequate long-term funding.

Another challenge is that patterns of disease and exposure vary over time and place, as illustrated in Table 7.7, which shows trends in breastfeeding over time in the U.K.

The challenge of variation in the prevalence of exposure and outcome can be tackled by employing a sequential design of cohorts distributed across time and space that collect comparable measures such as height and weight. It can be seen from Table 7.4 that a number of cohorts exist and that comparisons across cohorts are therefore possible.

Most birth cohorts rely on either self-reported or objective measures to identify obese children; thus, investigators have to deal with losses to follow-up as families drop out of the study. The percentage of child questionnaires returned from birth to age 9 years in ALSPAC is shown in Figure 7.1.

Such losses to follow-up decrease power and may introduce bias. A comprehensive approach to engaging the families can improve cohort retention; however, considerable losses to follow-up are inevitable in most birth cohorts. More sophisticated statistical approaches to modeling the effects of losses to follow-up, such as multiple imputation, can be used to improve power in longitudinal analyses where data are missing at random (Rubin, 1987). Sensitivity analyses can be used to investigate the plausibility of the missing at random assumption and to quantify the likely effects of potential bias.

Birth cohorts are better able to adjust for confounding than other observational studies. Like all cohorts, they are prone to confounding, particularly for behaviors such as obesity, which are socially patterned. Statistical adjustment for confounding in multivariable models may not fully correct for the effects of such social stratification. While the results of randomized double-blind controlled trials are not confounded, not all interventions can be tested experimentally. Where possible, investigators should look across birth cohort studies with different confounding structures to see whether the same exposure-obesity associations are observed. One problem is that exposure-obesity associations may vary over time and be either qualitatively or quantitatively different in a more obesogenic environment. For example, the risk of childhood

Table 7.6   Challenges for Birth Cohorts

Challenge	Description	Possible Solutions
Time and expense	Birth cohorts (like all cohorts) are expensive and take time for outcome data to accrue; therefore, they are difficult to fund with project grants	Program funding of cohorts as resources
Cohorts restricted by time and place of recruitment.	Patterns of exposure and outcome vary over time	A sequential design with repeated birth cohorts and comparative analyses in cohorts recruited in the past or recruited in other countries with different patterns of exposure and outcome
Cohort attrition	The numbers of participants still completing questionnaires or requiring health checks fall with time and are also patchy, with participants missing some data collection points; this can create problems with loss of power and bias	Skilled and committed fieldwork team Statistical imputation of missing data to increase power Sensitivity analyses to quantify the effect of any potential bias on associations
Confounding	Cohorts, like all observational studies, are prone to confounding, particularly for socially patterned outcomes	Where possible, key associations should be confirmed in randomized controlled trials The ecological explanatory power of associations should be used as a further check The presence of associations should be confirmed in populations with different confounding structures

Table 7.7 Proportions of Mothers Still Breastfeeding at Various Ages, by Location in the U.K. and by Year

Survey location (year)[†]	Percent Breastfeeding		
	At 1 month	At 3 months	At 6 months
London (1905–1919)[a]	90	80	70
Salford (1908–1913)[b]	89	83	78
Stafford (1918–1919)[b]	80	56	38
Aberdeen (1921–1922)[c]	—	—	50
Bristol (1929–1930)[d]	86	77	—
Birmingham (1938)[e]	87	50	—
Newcastle (1938)[f]		58	35
Edinburgh (1938)[f]		55	39
Luton (1945)[g]	80	57	39
England and Scotland (1946)[h]	60	42	30
Newcastle (1947–1948)[i]	78	48	31
Bristol (1947–1948)[d]	67	45	32
Luton (1956)[g]	70	38	15
Nottingham (1959–1960)[i]	54	29	13
Scotland (1970)[j]	15		
Dudley (1972)[k]	14	6	—
Cheltenham (1973)[l]	22*	15	—
Cambridge (1978)[m]	—	46	—
UK (1990)[n]	39*	—	21
UK (1995)[n]	42*	—	21

Note: Data are ordered by survey year.

*At 6 weeks.

Sources: [a]Fildes, V., *J. Biosoc. Sci.,* 24, 53–70, 1992; [b]Fildes, V., *Continuity Change,* 13, 251–280, 1998. [c]Baxter-Jones, A.D.G., et al., *Arch. Dis. Child.,* 81, 5–9, 1999; [d]Ross, A.I., *Lancet,* i, 630–632, 1951; [e]Ross, A.I., *Lancet,* i, 437–438, 1943; [f]Spence, J.C., *BMJ,* ii, 729–733, 1938; [g]Dykes, R.M., *Lancet,* ii, 230–232, 1957; [h]Douglas, J.W.B., *J. Obstetr. Gynaecol. Br. Empire,* 57, 335–361, 1950; [i]Newson, J. and Newson, E., *Infant Care in an Urban Community,* George Allen and Unwin Ltd., London, 1963. [j]Hall, B., *Lancet,* i, 779–781, 1975; [k]Shukla, A., et al., *BMJ,* iv, 507–515, 1972; [l]Department of Health and Social Security, *Present Day Practice in Infant Feeding,* HMSO, London, 1974; [m]Whichelow, M.J. and King, B.E., *Arch. Dis. Child.,* 54, 245, 1979; [n]Foster, K., et al., *Infant Feeding 1995: A Survey of Infant Feeding Practices in the United Kingdom,* The Stationery Office, London, 1997.

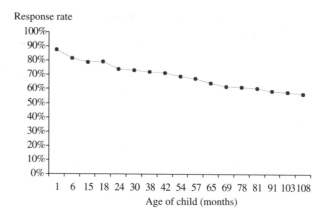

Figure 7.1 The Proportion of Child Questionnaires Returned From Birth to Age 9 Years in the Avon Longitudinal Study of Parents and Children (ALSPAC). [From The Avon Longitudinal Study of Parents and Children (ALSPAC)].

obesity conferred by a family history of obesity may be different in the current obesity epidemic than it was in the past.

Investigators should also look to see whether differences in exposure levels over time and geography explain differences in childhood obesity as individual based exposure-obesity associations, which have limited ecological explanatory power, are likely to be less important. For example, although some studies show associations between breastfeeding and subsequent obesity risk, it seems unlikely that the trends in breastfeeding in the U.K., shown in Table 7.7, explain the current epidemic of childhood obesity.

CONCLUSIONS

Childhood obesity is a serious public health problem. Action to prevent childhood obesity should be based on evidence. The results from birth cohorts and comparative analyses across birth cohorts can provide evidence on the causes of childhood obesity to help guide preventive efforts.

ACKNOWLEDGMENTS

I thank Jonathan Wells and John Reilly, whose ideas and work have helped to shape this chapter. I also thank Richard Martin who provided me with Table 7.7 and Jon Heron who provided me with Figure 7.1. Finally, I thank Jean Golding and the rest of the ALSPAC Study Team without whom there would be no study and no data to analyze and discuss.

REFERENCES

Andersen, R.E., Crespo, C.J., Bartlett, S.J., Cheskin, L.J., and Pratt, S.J., Relationship of physical activity and television watching with body weight and level of fatness among children. Results from the third National Health and Nutrition Examination Survey, *JAMA*, 279, 938–942, 1998.

Armstrong, J. and Reilly, J.J., Breastfeeding and lowering the risk of childhood obesity, *Lancet*, 359, 2003–2004, 2002.

Baxter-Jones, A.D.G., Cardy, A.H., Helms, P.J., Phillips, D.O., Smith, W.C., Influence of socioeconomic conditions on growth in infancy: the 1921 Aberdeen birth cohort. *Arch. Dis. Child.*, 81, 5–9, 1999.

Berenson, G.A., Srinivasan, S.R., Bao, W., Newman, W.P.I., Tracy, R.E., and Wattigney, W.A., Association between multiple cardiovascular risk factors and atherosclerosis in children and young adults, *N. Engl. J. Med.*, 338, 1650–1656, 1998.

Black, N. 1996, Why we need observational studies to evaluate the effectiveness of health care, *BMJ*, 312, 1215–1218, 1996.

Bundred, P., Kitchiner, D., and Buchan, I., Prevalence of overweight and obese children between 1989 and 1998: population based series of cross sectional studies, *BMJ*, 322, 326–328, 2001.

Chinn, S. and Rona, R.J., Can the increase in body mass index explain the rising trend in asthma in children?, *Thorax*, 56, 845–850, 2001.

Cooper, A.R., Page, A., Fox, K.R., and Misson, J., Physical activity patterns in normal, overweight and obese individuals using minute-by-minute accelerometry, *Eur. J. Clin. Nutr.*, 54, 887–894, 2000.

Davies, P.S.W., Diet composition and body mass index in pre-school children, *Eur. J. Clin. Nutr.*, 51, 443–448, 1997.

Department of Health and Social Security, *Present Day Practice in Infant Feeding*, HMSO, London, 1974.

Després, J.P., Lemieux, I., and Prud'homme, D., Treatment of obesity: need to focus on high risk abdominally obese patients, *BMJ* 322, 716–720, 2001.

Dewit, O., Fuller, N.J., Fewtrell, M.S., Elia, M., and Wells, J.C., Whole body air displacement plethysmography compared with hydrodensitometry for body composition analysis, *Arch. Dis. Child.*, 82, 159–164, 2000.

Dietz, W.H., Critical periods in childhood for the development of obesity, *Am. J. Clin. Nutr.*, 59, 955–959, 1994.

Dietz, W.H., Adiposity rebound: reality or epiphenomenon?, *Lancet*, 356, 2027–2028, 2000.

Dietz, W.H. and Gortmaker, S.L., Do we fatten our children at the television set? Obesity and television viewing in children and adolescents, *Pediatrics*, 337, 807–812, 1985.

Dietz, W.H. and Gortmaker, S.L., TV or not TV: fat is the question, *Pediatrics*, 91, 499–501, 1993.

Dietz, W.H.and Gortmaker, S.L., Preventing obesity in children and adolescents, *Annu. Rev. Public Health*, 22, 337–353, 2001.

Douglas, J.W.B., The extent of breast feeding in Great Britain in 1946, with special reference to the health and survival of children, *J. Obstetr. Gynaecol. Br. Empire*, 57, 335–361, 1950.

Drake, A.J., Smith, A., Betts, P.R., Crowne, E.C., and Shield, J.P.H., Type 2 diabetes in obese white children, *Arch. Dis. Child.*, 86, 207–208, 2002.

Dykes, R.M., Vitamin supplementation and type of feeding in infancy, *Lancet*, ii, 230–232, 1957.

Ebbeling, C.B., Pawlak, D.B., and Ludwig, D.S., Childhood obesity: public-health crisis, common sense cure, *Lancet,* 360, 473–481, 2002.

Ellis, K.J., Human body composition: in vivo methods, *Physiol. Rev.,* 80, 649–680, 2000.

Fagot-Campagna, A., Pettitt, D.J., Engelgau, M.M., Burrows, N.R., Geiss, L.S., Valdez, R., Beckles, G.L., Saaddine, J., Gregg, E.W., Williamson, D.F., and Narayan, K.M., Type 2 diabetes among North American children and adolescents: an epidemiologic review and a public health perspective. *J. Pediatr.,* 136, 664–672, 2000.

Ferri E., Forty years on: Professor Nevill Butler and the British Birth Cohort Studies. *Paediatr. Perinat. Epidemiol.* 12S, 31–44, 1998.

Figueroa-Munoz, J.I., Chinn, S., and Rona, R.J., Association between obesity and asthma in 4–11 year old children in the UK, *Thorax,* 56, 133–137, 2001.

Fildes, V., Breast-feeding in London, 1905–19, *J. Biosoc. Sci.* 24, 53–70, 1992.

Fildes, V., Infant feeding practices and infant mortality in England, 1900–1919, *Continuity Change,* 13, 251–80, 1998.

Foster, K., Lader, D., and Cheesborough, S., *Infant Feeding 1995: A Survey of Infant Feeding Practices in the United Kingdom;* Carried out by the Social Survey Division of ONS on behalf of the Department of Health, the Scottish Office Department of Health, the Welsh Office and the Department of Health and Social Services in Northern Ireland, The Stationery Office, London, 1997.

Freedman, D.S., Serdula, M.K., Srinivasan, S.R., and Berenson, G.A., Relation of circumference and skinfold thicknesses to lipid and insulin concentrations in children and adolescents: the Bogalusa Heart Study, *Am. J. Clin. Nutr.,* 69, 308– 317, 1999.

French, S.A., Story, M., and Jeffery, R.W., Environmental influences on eating and physical activity, *Ann. Rev. Public Health,* 22, 309–335, 2001.

Gale, C.R., Martyn, C.N., Kellingray, S., Eastell, R., and Cooper, C., Intrauterine programming of adult body composition, *J. Clin. Endocrinol. Metab.,* 86, 267–272, 2001.

Gale, E.A., Is there really an epidemic of type 2 diabetes?, *Lancet,* 362, 503–504, 2003.

Garn, S.M., LaVelle, M., Rosenberg, K.R., and Hawthorne, V.M., Maturational timing as a factor in female fatness and obesity, *Am. J. Clin. Nutr.,* 43, 879–883, 1986.

Garrow, J.S., Obesity, in *Human Nutrition and Dietetics,* 10th ed., Garrow, J.S., James, W.P.T., and Ralph, A., Eds., Churchill Livingstone, London, pp. 527–545, 2000.

Gillman, M.W., Rich–Edwards, J.W., Rifas-Shiman, S.L., Lieberman, E.S., Kleinman, K.P., and Lipshultz, S.E., Maternal age and other predictors of newborn blood pressure, *J. Paediatr.,* 144, 240–245, 2004.

Golding, J., Pembrey, M., Jones, R., and The ALSPAC Study Team, ALSPAC — The Avon Longitudinal Study of Parents and Children. I. Study methodology, *Paediatr. Perinat. Epidemiol.,* 15, 74–87, 2001.

Gunnell, D.J., Frankel, S.J., Nanchahal, K., Peters, T.J., and Davey Smith, G., Childhood obesity and adult cardiovascular mortality: a 57-y follow-up study based on the Boyd Orr cohort, *Am. J. Clin. Nutr.,* 67, 1111–1118, 1998.

Guo, S.S. and Chumlea, W.C., Tracking of body mass index in children in relation to overweight in adulthood, *Am. J. Clin. Nutr.,* 70, 145S–148S, 1999.

Hall, B., Changing composition of human milk and early development of appetite control, *Lancet,* i, 779–781, 1975.

Heitmann, L., Erikson, H., Ellsinger, B.M., Mikkelsen, K.L., and Larsson, B., Mortality associated with body fat, fat-free mass and body mass index among 60-year-old Swedish men — a 22-year follow-up. The study of men born in 1913, *Int. J. Obes.,* 24, 33–37, 2000.

Kimm, S.Y., Barton, B.A., Obarzanek, E., McMahon, R.P., Sabry, Z.I., Waclawiw, M.A., Schreiber, G.B., Morrison, J.A., Similo, S., and Daniels, S.R., Racial divergence in adiposity during adolescence: The NHLBI Growth and Health Study, *Pediatrics*, 107, E34, 2001.

Li, S., Chen, W., Srinivasan, S.R., Bond, M.G., Tang, R., Urbina, E.M., and Bernson, G.S., Childhood cardiovascular risk factors and carotid vascular changes in adulthood. The Bogalusa Heart Study, *JAMA*, 290, 2271–2276, 2003.

Livingstone, M.B.E., Childhood obesity in Europe: a growing concern, *Public Health Nutr.*, 4, 109–116, 2001.

Lobstein, T.J., James, W.P.T., and Cole, T.J., Increasing levels of excess weight among children in England, *Int. J. Obes. Relat. Metab. Disord.*, 27, 1136–1138, 2003.

Ludwig, D.S., Peterson, K.E., and Gortmaker, S.L., Relation between consumption of sugar-sweetened drinks and childhood obesity: a prospective, observational analysis, *Lancet*, 357, 505–508, 2001.

Magarey, A.M., Daniels, L.A., and Boulton, T.J.C., Prevalence of overweight and obesity in Australian children and adolescents: reassessment of 1985 and 1995 data against new standard international definitions, *Med. J. Aust.*, 174, 561–564, 2001.

Maynard, L.M., Wisemandle, W., Roche, A.F., Chumlea, W.C., Guo, S.S., and Siervogel, R.M., Childhood body composition in relation to body mass index, *Pediatrics*, 107, 344–350, 2001.

McGill, H.C. and McMahon, C.A. 2003, Starting earlier to prevent heart disease, *JAMA*, 290, 2320–2322, 2003.

Moore, M.S., Interactions between physical activity and diet in the regulation of body weight, *Proc. Nutr. Soc.*, 59, 193–198, 2000.

Must, A., Naumova, E., Phillips, S., and Rand, W.M., Menarcheal timing and weight status in relation to weight status in late adolescence, *Paediatr. Perinat. Epidemiol.*, 13, p. A18, 1999.

Must, A. and Strauss, R.S., Risks and consequences of childhood and adolescent obesity, *Int. J. Obes. Relat. Metab. Disord. Suppl.*, 23, S2–S11, 1999.

Must, A., Morbidity and mortality associated with elevated body weight in children and adolescents, *Am. J. Clin. Nutr. Suppl.*, 63, 445S–447S, 1996.

Must, A., Jacques, P.F., Dallal, G.E., Bajema, C.J., and Dietz, W.H. 1992, Long term morbidity and mortality of overweight adolescents: a follow-up of the Harvard Growth Study of 1922 to 1935, *N. Engl. J. Med.*, 327, 1350–1355, 1992.

National Task Force on the Prevention and Treatment of Obesity, Overweight, obesity, and health risk, *Arch. Int. Med.*, 160, 898–904, 2000.

Newman, W.P.I., Freedman, D.S., Voors, A.W., Gard, P.D., Srinivasan, S.R., Cresanta, J.L., Williamson, G.D., Webber, L.S., and Berenson, G.A., Relation of serum lipoprotein levels and systolic blood pressure to early atherosclerosis. The Bogalusa Heart Study, *N. Engl. J. Med.*, 314, 138–144, 1986.

Newson, J. and Newson E., *Infant Care in an Urban Community*, George Allen and Unwin Ltd., London, 1963.

Nieto, F.J., Szklo, M., and Comstock, G.W., Childhood weight and growth rate as predictors of adult mortality, *Am. J. Epidemiol.*, 136, 201–213, 1992.

Ogden, C.L., Troiano, R.P., Briefel, R.R., Kuczmarski, R.J., Flegal, K.M., and Johnson, C.L., Prevalence of overweight among preschool children in the United States, 1971 through 1994, *Pediatrics*, 99, 1–7, 1997.

Okasha, M., McCarron, P., McEwen, J., and Smith, G.D., Age at menarche: secular trends and association with adult anthropometric measures, *Ann. Hum. Biol.*, 28, 68–78, 2001.

Oken, E. and Gillman, M.W., Fetal origins of obesity, *Obes. Res.*, 11, 496–506, 2003.

Olsen, J., Melbye, M., Oslen, S.F., Sorensen, T.I., Aaby, P., Andersen, A.M., et al., The Danish National Birth Cohort: its background, structure and aim, *Scand. J. Public Health*, 29, 300–307, 2001.

Ong, K.K.L., Ahmed, M.L., Emmett, P.M., Preece, M.A., Dunger, D.B., and The ALSPAC Study Team, Association between postnatal catch-up growth and obesity in childhood: prospective cohort study, *BMJ*, 320, 967–971, 2000.

Parsons, T.J., Power, C., Logan, S., and Summerbell, C.D.1999, Childhood Predictors of adult obesity: a systematic review, *Int. J. Obes. Relat. Metab. Disord.*, 23 (Suppl. 8), S1–S107, 1999.

Parsons, T.J., Power, C., and Manor, O., Fetal and early life growth and body mass index from birth to early adulthood in 1958 British cohort: longitudinal study, *BMJ*, 323, 1331–1335, 2001.

Parsons, T.J., Power, C., and Manor, O., Infant feeding and obesity through the lifecourse, *Arch. Dis. Child.*, 88, 793–794, 2003.

Power, C., Lake, J.K., and Cole, T.J., Body mass index and height from childhood to adulthood in the 1958 British born cohort, *Am. J. Clin. Nutr.*, 66, 1094–1101, 1997.

Power, C. and Jefferis, J.M.H., Fetal environment and subsequent obesity: a study of maternal smoking, *Int. J. Epidemiol.*, 31, 413–419, 2002.

Prentice, A.M. and Jebb, S.A. 1995, Obesity in Britain: gluttony or sloth?, *BMJ*, 311, 437–439, 1995.

Raitakari, O.T., Juonala, M., Kähönen, M., Taittonen, L., Laitinen, T., Mäki-Torkko, N., Järvisalo, M.J., Uhari, M., Jokinen, E., Rönnemaa, T., Åkerblom, H.K., and Viikari, J.S.A., Cardiovascular risk factors in childhood and carotid artery intima-media thickness in adulthood. The Cardiovascular Risk in Young Finns Study, *JAMA*, 290, 2277–2283, 2003.

Ravelli, G., Stein, Z.A., and Susser, M., Obesity in young men after famine exposure in utero and early infancy, *N. Engl. J. Med.*, 295, 349–353, 1976.

Reilly, J.J. and Dorosty, A.R., Epidemic of obesity in UK children, *Lancet*, 354, 1874–1875, 1999.

Reilly, J.J., Dorosty, A.R., and Emmett, P.M., Prevalence of overweight and obesity in British children: cohort study, *BMJ*, 319, 1039, 1999.

Reilly, J.J., Dorosty, A.R., Emmett, P.M., and The ALSPAC Study Team, Identification of the obese child: adequacy of the body mass index for clinical practice and epidemiology, *Int. J. Obes. Relat. Metab. Disord.*, 24, 1623–1627, 2000.

Report of a Joint WHO/FAO Expert Consultation, *Diet, Nutrition and the Prevention of Chronic Diseases*, World Health Organization, Geneva, p. 916, 2003.

Richter L.M., Norris, S.A., and de Wet T., Transition from birth to ten to birth to twenty: the South African Cohort reaches 13 years of age, *Paediatr. Perinat. Epidemiol.*, 18, 290–301, 2004.

Robinson, T.N., Reducing children's television viewing to prevent obesity: a randomized controlled trial, *JAMA*, 282, 1561–1567, 1999.

Rolland-Cachera, M.F., Deheeger, M., Bellisle, F., Sempe, M., Guilloud-Bataille, M., and Patois, E., Adiposity rebound in children: a simple indicator for predicting obesity, *Am. J. Clin. Nutr.*, 39, 129–135, 1994.

Ross, A.I., Breast-feeding in Bristol, *Lancet*, i, 630–632, 1951.

Ross, A.I., Is breastfeeding worthwhile?, *Lancet*, i, 437–438, 1943.

Rubin, D., *Multiple Imputation for Non-Response in Surveys* Wiley, New York, 1987.

Rudolf, M.C.J., Sahota, P., Barth, J.H., and Walker, J., Increasing prevalence of obesity in primary school children: cohort study, *BMJ,* 322, 1094–1095, 2001a.

Rudolf, M.C.J., Sahota, P., Barth, J.H., and Walker, J., Increasing prevalence of obesity in primary school children: cohort study, *BMJ,* 322, 1094–1095, 2001b.

Shukla, A., Forsyth, H.A., Anderson, C.M., and Marwah, S.M., Infantile overnutrition in the first year of life: a field study in Dudley, Worcestershire, *BMJ,* iv, 507–15, 1972.

Sorensen, T.I.A. and Echwald, S.M., Obesity genes, *BMJ,* 322, 630–631, 2001.

Sorensen, T.I.A., Holst, C., Stunkard, A.J., and Skovgaard, L.T., Correlations of body mass index of adult adoptees and their biological and adoptive relatives, *Int. J. Obes. Relat. Metab. Disord.,* 116, 227–236, 1992.

Spence, J.C., The modern decline of breast feeding, *BMJ,* ii, 729–33, 1938.

Stunkard, A.J., Foch, T.T., and Hrubec, Z., A twin study of human obesity, *JAMA,* 256, 51–54, 1986.

van Lenthe, F.J., Kemper, C.G., and van Mechelen, W., Rapid maturation in adolescence results in greater obesity in adulthood: the Amsterdam Growth and Health Study, *Am. J. Clin. Nutr.,* 64, 18–24, 1996.

Vik, T., Jacobsen, G., Vatten, L., and Bakketeig, L.S., Pre- and post-natal growth in children of women who smoked in pregnancy, *Early Hum. Dev.,* 45, 245–255, 1996.

Visscher, T.L. and Seidell, J.C., The public health impact of obesity, *Ann. Rev. Public Health,* 22, 355–375, 2001.

von Mutius, E., Schwartz, J., Neas, L.M., Dockery, D., and Weiss, S.T., Relation of body mass index to asthma and atopy in children: the National Health and Nutrition Examination Study III, *Thorax,* 56, 835–838, 2001.

Wabitsch, M., The acquisition of obesity: insights from cellular and genetic research, *Proc. Nutr. Soc.,* 59, 325–330, 2000.

Wadsworth, M. and Hardy, R., Coronary heart disease morbidity by age 53 years in relation to childhood risk factors in the 1946 birth cohort, in *National Heart Forum, A Lifecourse Approach to Coronary Heart Disease Prevention. Scientific and Policy Review,* Stationery Office, London, pp. 39–47, 2003

Weinsier, R.L., Hunter, G.R., Heini, A.F., Goran, M.I., and Sell, S.M., The etiology of obesity: relative contribution of metabolic factors, diet, and physical activity, *Am. J. Med.,* 105, 145–150, 1998.

Wells, J.C., Is obesity really due to high energy intake or low energy expenditure?, *Int. J. Obes. Relat. Metab. Disord.,* 22, 1139–1140, 1998.

Wells, J.C., Coward, W.A., Cole, T.J., and Davies, P.S., The contribution of fat and fat-free tissue to body mass index in contemporary children and the reference child, *Int. J. Obes. Relat. Metab. Disord.,* 26, 1323–1328, 2002.

Wells, J.C., Fuller, N.J., Dewit, O., Fewtrell, M.S., Elia, M., and Cole, T.J., Four-component model of body composition in children: density and hydration of fat-free mass and comparison with simpler models, *Am. J. Clin. Nutr.,* 69, 904–912, 1999.

Wells, JC.K., A Hattori chart analysis of body mass index in infants and children, *Int. J. Obes.,* 24, 325–329, 2000.

Whichelow, M.J. and King, B.E., Breast feeding and smoking, *Arch. Dis. Child.,* 54, 245, 1979.

Whitaker, R.C. and Dietz, W.H., Role of the prenatal environment in the development of obesity, *J. Pediatr.,* 132, 768–776, 1998.

Whitaker, R.C., Wright, J.A., Pepe, M.S., Siedel, K.D., and Dietz, W.H., Predicting obesity in young adulthood from childhood and parental obesity, *N. Engl. J. Med.*, 337, 869–873, 1997.

Williams, D.P., Going, S.B., Lohman, T.G., Harsha, D.W., Srinivasan, S.R., Webber, L.S., and Berenson, G.A., Body fatness and risk for elevated blood pressure, total cholesterol, and serum lipoprotein ratios in children and adolescents, *Am. J. Public Health*, 82, 358–363, 1992.

Wright, C.M., Parker, L., Lamont, D., and Craft, A.W., Implications of childhood obesity for adult health: findings from thousand families cohort study, *BMJ*, 323, 1280–1284, 2001.

8

BIOBEHAVIORAL DETERMINANTS OF ENERGY INTAKE AND CHILDHOOD OBESITY

M. B. E. Livingstone and K. L. Rennie

CONTENTS

INTRODUCTION

The burgeoning rates of childhood obesity in genetically stable populations suggest that an increasingly obesogenic environment is the major driving force behind this epidemic. Fostering the delicate balance between energy intake (EI) and energy expenditure is now exceedingly difficult. In the past few decades, several key environmental and cultural factors have converged to increase the probability of overeating. Although, within the scope of this chapter, only the EI side of the energy balance equation will be considered, it needs to be stressed that the overeating that leads to obesity is only a relative phenomenon. The current generation of children may be the most inactive of all time, and this factor is at least as important, if not more important, than overeating in accounting for the escalating rates of pediatric obesity. In the face of reduced energy needs, children are particularly vulnerable to passive overeating of food in the midst of a food environment that is less than benign. The factors that now conspire to undermine normal appetite regulation include the ready availability and variety of palatable, relatively low-cost, energy-dense foods served in ever-increasing portion sizes. This chapter evaluates these factors, along with other biobehavioral influences, such as eating frequency, that are all implicated in childhood obesity.

FOOD INTAKES

Although there is now a substantial body of research assessing the links between children's weight status and a number of dietary predictors of relevance, including total EI, percent EI from fat (% fat), and fat preferences, the outcomes of these studies are far from conclusive. Thus, whereas some cross-sectional data have demonstrated that higher weight status is correlated with higher total daily EI, the majority of studies have failed to demonstrate that fatter children have higher EI. Indeed, many studies have found that fatter children consume a lower EI relative to their weight than their lean counterparts.

More recently, a number of studies, the majority of which have been cross-sectional, have focused on the role of macronutrient intake, particularly fat intake, in promoting a positive energy balance (Gazzaniga and Burns, 1993; Obarzanek et al., 1994; Klesges et al., 1995; Ortega et al., 1995; Rolland-Cachera et al., 1995; Maffeis et al., 1998; Nguyen et al., 1996; Davies, 1997; Tucker et al., 1997; Guillaume et al., 1998; Robertson et al., 1999; Magarey et al., 2001; McGloin et al., 2002; Remer et al., 2002). However, once again, the outcomes of these studies are not easy to interpret.

Higher fat intake (% energy) has been concurrently associated with higher percentage body fat (Guillaume et al., 1998; Gazzaniga and Burns, 1993; Tucker et al., 1997), fat mass (Nguyen et al., 1996; McGloin et al., 2002), skinfold thickness (Obarzanek et al., 1994; Guillaume et al., 1998), and body-mass index (BMI) (Klesges et al., 1995; Ortega et al., 1995). In some studies, inverse associations with carbohydrate intake were shown (Obarzanek et al., 1994; Tucker et al., 1997; Guillaume et al., 1998; Magarey et al., 2001). On the other hand, a number of longitudinal studies have either failed to identify an independent association between children's % fat intake and their weight status (Rolland-Cachera et al., 1995; Davies, 1997; Maffeis et al., 1998; Remer et al., 2002) or detected an inconsistent relationship (Robertson et al, 1999; Magarey et al. 2001). Two longitudinal studies (Rolland-Cachera et al., 1995; Maffeis et al., 1998) found that protein intake (% energy) was a significant predictor of later adiposity in younger children; however, in older children, macronutrient intake was not predictive of body fatness. In children followed longitudinally for 2 to 15 years, there was no association between % fat intake and either BMI or total body fat (by triceps skinfold thickness), but a positive association with abdominal fat (by subscapular skinfold thickness) was observed (Magarey et al., 2001). In this latter study, macronutrient intake at an earlier age did not predict later body fatness. Rather, the best predictor of current body fatness was adiposity at an earlier age, together with parental adiposity.

Overall, no definitive conclusions can be drawn from these studies. Paradoxically, the prevalence of pediatric obesity in the U.K. continues to increase, despite a reported decrease in both total EI and % fat intake in U.K. children and adolescents (National Diet and Nutrition Survey, 2000). Falling EI are likely to be associated with the secular decline in physical activity levels, but there are a number of other reasons why the integrity of food intake data may be questioned. First, many of the studies concerned with assessing the associations between dietary intake and body composition are remarkable for their lack of consistency in the methods employed to estimate body fatness or nutrient intake. Second, discrepancies in the reported relationships between EI and weight status are likely to be due to unquantifiable errors in both self-reported food intake and body composition assessments. Bias in the reporting of children's food intakes could be selectively targeted at foods that are perceived to be "bad" or "unhealthy." Given the current health campaigns aimed at promoting low-fat diets, it could be surmised that the intake of high-fat foods is more vulnerable to underreporting, whereas the intake of low-fat foods may be exaggerated. Consequently, the apparent decrease in EI and dietary fat intake may be more apparent than real. Third, potential confounding factors such as the varying energy needs of children, as a consequence of growth rate and gender, have not always been accounted for. Finally, given that it takes only a small cumulative positive energy balance to ultimately result in obesity, it is highly debatable whether any of the existing instruments for assessing dietary intake and energy expenditure are sufficiently sensitive to detect such small energy imbalances at discrete time points.

On the other hand, more sensitive indices of future obesity risk may be the food and activity *preferences* shared by family members. It is highly unlikely that a child's food and activity preferences are isolated events within the family context; rather they are likely to reflect a constellation of characteristics shared by family members, which may or may not be conducive to obesity. However, relatively few studies have evaluated the relationship between food preferences, food intake, and weight status of both parents and their children. Fisher and Birch (1995) and Ricketts (1997) found that fatter children displayed a higher preference for fat, which in turn was a positive predictor of their fat intake. Moreover, Fisher and Birch (1995) observed that parental adiposity was a predictor of their child's preferences for and consumption of fat. Lean children at high risk of obesity (i.e., with obese/overweight parents) have also been found to exhibit higher preference for fatty foods, lower preference for vegetables, and a more externally reactive eating style compared with lean children at low risk of obesity (i.e., with lean parents). Thus the dietary dispositions of the high-risk children, compounded by their stated preferences for sedentary activity, could represent a particularly permissive environment for future obesity.

ENERGY DENSITY

The energy density (ED) of foods is a key determinant of overall EI and therefore of body weight regulation. If it is assumed that individuals tend to eat a constant volume of food each day, in theory at least, lowering the ED of the diet represents a promising approach to weight maintenance/ reduction strategies. However, because consumers' food choices are based largely on taste, cost, and convenience, it remains unclear whether humans in Western cultures would adapt to consuming a low-energy-dense diet in the longer term, particularly if it compromised the hedonic qualities of the diet. Aside from issues of compliance, it must be remembered that low-energy-dense diets, high in fruit and vegetables, are likely to incur higher food costs (expressed as kJ/£). Recommendations to follow such diets should be tempered with the caution because, for many families, these diets are simply not an affordable option.

The ED of individual foods is a function of their water content (negative predictor) and, to a lesser extent, their fat content (positive predictor); however, there are a number of caveats to the fat content–ED relationship. First, despite the reasonably good correlation between the fat content of foods and their ED, it is not inevitable that high-fat foods (expressed as % energy) have a high ED. For example, the ED of the diets of preschool children was positively associated not only with fat but also with a range of starchy foods, including breakfast cereals, bread, and potatoes (Gibson, 2000). Second, there is a considerable overlap between the ED of high-fat and high-carbohydrate foods, particularly in snacks. Third, consumption of reduced or fat-free foods does not lower the ED of the diet. It is not always appreciated that, in order to retain sensory appeal, many lower fat foods have a high refined carbohydrate and lower water content, giving them an ED approximating that of the high-fat-equivalent food. These caveats are important because fat reduction (and by implication ED reduction) messages have become the cornerstone of dietary advice to treat/prevent obesity. This in turn has resulted in the rapid proliferation of low-fat foods and, arguably, has fueled a fat phobia in the population. Paradoxically, this phobia may have helped to generate the misperception that, as long as low-fat or fat-free products are consumed, the quantity of food eaten is unimportant (American Institute for Cancer Research, 2000). Finally, it appears that energy provided in fluid form does not induce a compensatory reduction in food intake as efficiently as energy delivered in solid foods (Mattes, 1996; DiMeglio and Mattes, 2000). Although this finding does not discount the overriding importance of ED, it suggests that the physical state of foods may be important in EI regulation. It also implies that the ED of foods and beverages should be considered separately, although this is not usually the case.

Energy density appears to have a direct influence on EI, which is independent of its fat content (Stubbs et al., 1996; Rolls et al., 1999). However, the specific characteristics, concentrations, and relative proportions of the macronutrients, in addition to the physical form of foods, may also mediate the effects of ED. Thus, although it is widely accepted that the effects of fat in promoting overconsumption of energy is heavily influenced by its ED, nevertheless, the oro-sensory properties of fat, coupled with its low postabsorptive satiety value, may promote high fat hyperphagia that is independent of its contribution to dietary ED. However, although high-fat diets increase the probability of overeating, it would be misleading to conclude that high-fat diets are the sole cause of obesity. On the other hand, carbohydrates appear to exert a more acute effect on satiety than fats, and because, on average, carbohydrate foods have a lower ED than fat foods and often contain dietary fiber, the intake of high-carbohydrate foods is likely to be more self limiting. However, there is still considerable scope for high-carbohydrate foods to promote high levels of EI (Stubbs et al., 1998).

Undoubtedly, the high palatability of many high-energy-dense foods is also likely to favor their overconsumption. Drewnoski (1997) has provided compelling evidence that, whereas fats and sugars have distinct sensory attributes, in combination, they appear to have a synergistic effect and exert a particularly potent stimulus on EI. Unfortunately, although ED foods are palatable, they are not especially satiating; therefore, they are liable to be overeaten.

HOW WELL DO CHILDREN RESPOND
TO THE ED OF FOODS?

Children's food preferences and dietary choices are primarily driven by taste. They are born with an innate preference for sweet and salty tastes, an innate aversion to sour and bitter tastes, and a neophobic reaction to new foods. There is no evidence that children have an innate unlearned preference for high-fat, high-energy-dense foods, but they are predisposed to learn conditioned flavor preferences for these foods (Johnson et al., 1991; Birch and Fisher, 1998). Given a choice, young children prefer high-ED foods to those lower in ED (Johnson et al., 1991; Birch, 1992), and these preferences may be compounded by the association of high-energy-dense foods with special occasions or when such foods are offered as treats or rewards. A positive relationship between ED and food preferences has been observed in 4-year-old U.K. children: the best liked foods were either ED or sweet (Wardle et al., 2001b).

Given the powerful biological and environmental triggers to eat high-fat and sweet foods, how well do children self-regulate their food intake? Early studies suggest that, when presented with a range of healthy foods, infants and young children are sensitive to internal hunger and satiety cues and can effectively self-regulate their EI (Davis, 1928; Story and Brown, 1987). In a series of experiments conducted by Birch et al., it was also observed that preschool children can adjust the amount of food consumed in response to changes in ED both within (Birch and Deysher, 1985) and between meals (Birch et al., 1991). Moreover, when the ED of meals was diluted by substitution of fat with olestra, self-selected intake was increased in order to defend total 24-hr EI (Birch et al., 1993). However, as cautioned by Birch, it is uncertain whether this responsiveness to ED would be maintained in the face of a very high-fat diet. Indeed, there is evidence to suggest that, when children are given free access to a range of palatable foods following a meal where they have eaten to fullness, children will overeat even in the absence of hunger (Birch and Fisher, 2000).

This strongly suggests that children's ability to self-regulate their EI may be limited to those situations where food choices are heavily constrained and purposively directed toward "healthy" foods. To date, no research has been carried out in a naturalistic setting in which the eating behavior of children given unrestricted access to highly palatable food is observed.

With increasing age, factors other than hunger and satiety cues begin to influence food intake, resulting in decreased responsiveness to the ED of foods. For example, by the age of 9 to 10 years, U.K. children have reported that they had already dieted to lose weight, suggesting that food intake is driven more by cognitive than by physiological factors (Hill et al., 1992). There are also substantial individual differences between children in their ability to regulate food intake in response to its ED, implying that the response to ED is relatively fragile and could easily be disrupted by well-intended but inappropriate child-feeding practices that attempt to externally control what and how children eat.

CARBOHYDRATES

Views about carbohydrates in relation to obesity have oscillated considerably in recent decades. In the 1970s, carbohydrates, particularly sucrose, were regarded as the most important dietary factor predisposing to weight gain; however, by the 1990s, their role in obesity was considered to be more or less benign. There is now a reasonable body of evidence to support the conclusion that carbohydrates are more satiating than fat and that, on average, high-fat diets will tend to promote higher EI than low-fat (high-carbohydrate) diets (Blundell and Stubbs, 1999). However, it remains unclear whether all carbohydrates have similar effects on appetite, EI, and overall energy balance.

SUGAR

Currently, one of the most contentious issues in the diet and obesity debate is the role of sugar in promoting obesity. One of the notable changes in children's diets over the past few decades has been the increase in sugar-sweetened soft drinks, which now constitute the primary source of added sugars in the diet. Between 1965 and 1996, soft drink intake of U.S. adolescents increased by 187% and 123% in boys and girls, respectively, while milk intake declined by approximately 40% (Cavadini et al., 2000). In the U.K. National Diet and Nutrition Survey (2000), boys and girls derived 16.7% and 16.4%, respectively, of their food energy from nonmilk extrinsic sugars (NMES). Sugar-sweetened soft drinks were consumed by 75% of the survey group and, along with chocolate confectionery, were the main sources of NMES in the diet. Consumption of sugar-sweetened soft drinks has been positively associated with children's EI (Harnack et al., 1999; Mrdjenovic and Levitsky, 2003), and obese children have been shown to consume a greater proportion of their total EI from soft drinks compared with nonoverweight children (Troiano et al., 2000). Ludwig et al. (2001) concluded that consumption of sugar-sweetened soft drinks is an independent risk factor for obesity in children aged 11 to 12 years and that, for every additional serving of these drinks consumed, the odds of becoming obese increased by 60%. However, in this study, a large proportion of the children were already obese, and it is uncertain whether the observed relationship between the intake of nondiet soft drinks and adiposity would be evident across the range of BMI. A more recent study by Mrdjenovic and Levitsky (2003) suggests that this may be the case. Children with the highest intakes of sweetened soft drinks had the highest EI and gained the most weight during the study. Although it is difficult to prove causality from short-term observational studies, the results of the latter study are consistent with the observation that consumption of sugar-sweetened soft drinks could be permissive for obesity because of an imprecise and incomplete compensation for sugar consumed in liquid form (Mattes, 1996; DiMeglio and Mattes, 2000).

As a result of these observations, the WHO has recommended that adults and children limit their consumption of free sugars, particularly sugar-sweetened soft drinks to less than 10% of total energy in order to prevent obesity (WHO, 2003). More recently, the American Academy of Pediatrics has advocated that sweetened drinks should no longer be available in U.S. schools (American Academy of Pediatrics, 2004). Given the implications of these recommendations for the food industry, particularly the sugar industry, the debate about sugar and obesity is unlikely to subside for the foreseeable future.

These findings and recommendations remain controversial since studies suggest that free sugars in the diet have either no effect in promoting adiposity or are negatively correlated with body weight (Astrup and Raben, 1995; Hill and Prentice, 1995; Gibson, 1996) This has led to the suggestion that a high intake of sucrose may help prevent weight gain by displacing fat from the diet. However, given the pervasive problem of biased reporting in dietary surveys, particularly among the obese, it is virtually impossible to make valid conclusions about associations between sugars and body weight from self-reported dietary data. Nevertheless, these studies have probably helped to generate an oversimplistic, even misleading, perception that carbohydrates, especially sugar, will prevent weight gain and conversely that high fat intakes are the overriding risk factor for obesity. In reality, fat and sugars are generally not consumed in isolation, and both animals and humans tend to show strong preferences for mixtures of the two. Although not proven, it could be that sweet, high-fat foods are more potent at upregulating EI than foods rich in either sugars or fats. (Drewnoski, 1997).

SUGARS VS. COMPLEX CARBOHYDRATES

Studies on the comparative effects of different monosaccharides and polysaccharides on food intake have yielded inconclusive results. Some of the discrepancies can be attributed to the substantial differences in methodological approaches in these studies as well as the problems of controlling for the confounding effects. After a 14-day *ad libitum* intake of either high-sucrose, high-starch, or high-fat diets, Raben et al. (1997) observed a significant reduction in EI for individuals on the high-starch diet but no changes for individuals on either the high-fat or high-sucrose diets. The intake restraining effects of the high-starch diet were attributed to the higher volume and fiber content as well as the lower palatability of this diet. In addition to the increased palatability of the high-sugar diets, the higher consumption of sugar-containing drinks by the subjects may have contributed to the higher EI.

On the other hand, results of the CARMEN study have suggested that after a 6-month *ad libitum* intake of low-fat diets high in either sugar or complex carbohydrate, the effects on body weight regulation are indistinguishable (Saris et al., 2000). In both cases, spontaneous weight reduction was significantly greater compared with that seen on the normal fat diet, but it should be emphasized that the effects of the high-carbohydrate/low-fat diets were probably mediated through their lower ED. Although the effects of different classes of carbohydrates on long-term energy balance require confirmation, the results of this study underscore the importance of a reduction in overall ED of the diet for combating obesity. However, although bulky, high-fiber, low-ED diets may facilitate spontaneous reductions in EI, there is a caveat to any advice aimed at promoting such diets. Fiber intakes of Western diets remain stubbornly low, probably because high-fiber foods are not palatable to most people in the longer term.

GLYCEMIC INDEX OF CARBOHYDRATES

At present, the possible beneficial role of a low glycemic index (GI) diet in curbing passive overconsumption of energy and assisting long-term weight regulation is another "hot" dietary issue in obesity research. Glycemic index is defined as the incremental positive area under the blood glucose response curve after consumption of 50 g of carbohydrate from a test food relative to the effect of consuming 50 g of either white bread or glucose (reference food). In general, high-GI foods are characterized by their higher refined carbohydrate content, high-glucose and/or starch content relative to lactose, sucrose and fructose content, and low soluble fiber content.

The pros and cons of GI on appetite, EI, and weight regulation have recently been vigorously defended by Pawlak et al. (2002) and Raben (2002), respectively. On the one hand, Pawlak et al. (2002) asserted that consumption of high-GI foods, relative to low-GI foods, triggers a sequence of hormonal events that stimulates the rapid return of hunger sensations, promotes excessive EI, and possibly increases the risk of central adiposity, cardiovascular disease and non-insulin-dependent diabetes. On the other hand, Raben (2002), after reviewing 31 acute intervention trials, found that low-GI vs. high-GI meals incited greater satiety and reduced hunger in only half of the studies, with the remainder of the studies demonstrating either no differences in satiety ($n = 14$) or enhanced satiety following high-GI meals ($n = 2$). Among 20 longer-term intervention trials (<6 months), only 4 studies were able to report a greater weight loss following a low-GI diet, whereas 14 studies found no difference. On the basis of this analysis, Raben (2002) justifiably concluded that, at present, there was too little evidence to advocate the supremacy of a low-GI over a high-GI diet for long-term weight regulation.

Only a limited number of studies have investigated the impact of a low-GI diet on children's food intakes and weight regulation, with equally inconclusive findings. A high-GI vs. low-GI breakfast was found to stimulate hunger and elevate subsequent *ad libitum* energy intake by 53% in obese adolescents (Ludwig et al., 1999). On procedural grounds, the outcomes and implications of some studies are difficult to interpret. Thus, although a diminution in energy intake at lunchtime following a low-GI breakfast has recently been observed in both lean and overweight children aged 9 to 15 years, the significance of these results are difficult to assess because of the variations in the macronutrient and fiber contents across the test breakfasts (Warren et al., 2003). In an out-patient setting, Spieth et al. (2000) compared the impact of either an energy-reduced diet based on current healthy eating guidelines or a non-energy-reduced diet based on low-GI foods in the management of pediatric obesity. Although reductions in body weight were greater in the low-GI group, once again, interpretation of the results is difficult because a number of dietary variables, other than GI, were different in the two groups. Taken together, no clear pattern emerges about the possible protective role of a low-GI diet on appetite and weight regulation in children and adolescents. On one issue, the protagonists in this debate agree. The issue will not be resolved until sufficiently powered, randomized long-term control trials are undertaken in which *ad libitum* food intakes and changes in body composition are monitored following diets similar in all respects except GI. Nevertheless, despite the lack of compelling evidence, a low-GI diet for the treatment and prevention of obesity has now been advocated by some researchers (Ludwig, 2000; Roberts, 2000; Pawlak et al., 2002). Whether such advice can be justified at present is a matter of debate.

In conclusion, although a reasonable amount of evidence now exists to support the position that, on average, high-carbohydrate diets will tend to constrain EI relative to high-fat diets, the perception that all carbohydrates are protective against weight gain is less securely based than previously assumed. Adding carbohydrate, and especially sugar, to the diet cannot be assumed to displace fat or protect against excessive EI (Stubbs et al., 2001). To date, there seem to be few differences between sugars and starches on satiety and EI, but the effect of different types of carbohydrates (sugars vs. starches; high GI vs. low GI) requires further investigation. Further studies are also urgently required to validate the hypothesis that carbohydrates from fluids may be particularly conducive to weight gain.

EATING PATTERNS

The past three decades have witnessed enormous changes in the eating patterns and lifestyles of U.K. families. The shifts in dietary patterns that are most often cited as promoting pediatric obesity include an increase in out-of-home eating, particularly fast food consumption, increased snacking, and larger portion sizes of food. However, although it is reasonable to assume that these trends are causally related to obesity, there is surprisingly little hard evidence to support such a relationship. What evidence does exist is mainly cross-sectional in nature, based on relatively small sample sizes, and specific to the U.S. situation. Moreover, epidemiological studies that attempt to analyze the association between eating patterns, dietary quality, and obesity causality are inevitably hampered by the highly interrelated nature of the dietary exposures. Thus, it is exceedingly difficult, if not impossible, to distinguish the independent effects of nutrients, foods, or specific eating patterns on weight status.

Fast Food Consumption

Changing demographics mean that children across all income strata are spending a greater part of their day away from home, and, as a result, out-of-home eating is having an increasingly important influence on the nutritional quality of children's diets. Cross-sectional data from U.S. national food consumption surveys carried out during the 1970s, 1980s, and 1990s have shown a marked decline in in-home EI in favor of an increasing proportion of energy consumed in restaurants and fast food outlets. It is estimated that the latter has increased from 6.5% in 1978 to 19.3% in 1996 (Jahns et al., 2001; Nielson et al., 2002).

The appeal of fast food to children lies in its easy accessibility, convenience, consistency in product quality and menu offerings, and relatively low cost. On the other hand, the high palatability, high ED, and "value-for-money" large portion sizes of typical fast foods can easily incite passive overconsumption of energy. It has been estimated that, on average, the ED of fast foods is about 1.5 times greater than the average ED of the diets of British girls (Prentice and Jebb, 2003). Thus, regular consumption of such foods could easily undermine normal appetite and satiety mechanisms, particularly in children who are sedentary. Although there is some evidence to suggest that overall EI are greater among adolescents who regularly consume fast food relative to those who do not (French et al., 2001), there are no comparable data on children. Moreover, regular consumption of fast food in adolescents may promote greater intakes of soft drinks as well as compromise the intakes of fruit, vegetables, milk, and grain products (French et al., 2001).

Meal Frequency

It has been hypothesized that less structured eating patterns, particularly an increase in snacking behavior, is contributing to the increase in obesity (Drummond et al., 1996). Although intuitively plausible, this has been difficult to substantiate because the distinction between meals and snacks has blurred over time (Gatenby, 1997). Evidence suggests that adolescents with a consistent eating pattern based on three meals per day may be leaner than those with a more erratic eating pattern (Siega-Riz et al., 1998), an observation that concurs with studies showing a link between obesity and meal skipping (Bellisle et al., 1988; Siega-Riz et al., 1998). Conversely, an inverse cross-sectional relationship between the frequency of eating and body weight has also been noted, suggesting that a "grazing" eating pattern may be protective against weight gain (Kant et al., 1995; Summerbell et al., 1996). However, these latter studies are particularly prone to the generation of spurious relationships due to dietary misreporting and post hoc changes in eating patterns as a result of weight gain.

The snacking patterns of children and adolescents have changed dramatically over the past two decades. More children snack than ever before, with the greatest increase seen in the last decade (Jahns et al., 2001). The average snack size and energy content appears to have remained relatively stable over time, but, given that the number of snacking occasions has increased significantly, the overall impact has been a steady increase in the proportion of daily EI supplied by snacks. Of particular relevance to obesity is the observation that the ED of the nonsnacking component of the diet has remained constant, whereas the ED of snacks has increased (Jahns et al., 2001). The latter finding is of concern because even small increases in the ED of foods consumed can lead to large increases in EI.

Portion Sizes

Only recently has interest focused on the contribution that larger portion sizes may be making to the overconsumption of energy and the increasing prevalence of obesity. The trend toward increasing portion sizes, which started in the late 1970s, has been accelerating ever since. As a result, most current portions of restaurant food in the U.S. exceed, often significantly so, the recommended portion sizes (Young and Nestle, 2002). Of interest, the portion sizes of fast foods in the U.K. lag somewhat behind comparable foods from the same fast food chains in the US ... but for how long? More extensive analyses of nationally representative dietary data in the U.S. have also confirmed the trend toward larger portion sizes, not only for out-of-home eating but also for in-home consumption. Under controlled laboratory conditions, portion size has been demonstrated to positively influence EI in adults (Edelman et al., 1986; Rolls et al., 2002). For example, Rolls et al. (2002) showed that both lean and overweight subjects consumed 30% more energy when the serving size of macaroni and cheese was doubled from 500 g to 1000 g. Not only were the subjects unaware of the change in the amount of food served, but they did not feel any fuller as a result of eating the larger portion size.

In early childhood, it appears that children can effectively self-regulate their EI both within and between meals (Birch et al., 1991). Over a 20-year period, the temporal stability in average portion sizes consumed by children aged 1 to 2 years suggests that their eating behavior is primarily driven by internal hunger and satiety cues (McConahy et al., 2002) However, by the age of 3 to 4 years, eating behavior becomes much more reactive to environmental stimuli, implying that the association between body weight and portion size starts at an early age. Thus, although 3-year-old children were found to exhibit a consistent response by eating the same amount of macaroni and cheese irrespective of the portion size presented, 5-year old children responded immediately by consuming significantly more energy when presented with a large vs. small portion (Rolls et al., 2000). Furthermore, it appears that this is not a transient response that is compensated for in subsequent meals. Rather, the effect tends to persist on repeated exposure to large portion sizes (Fisher et al., 2003). In addition, in contrast to the earlier study (Rolls et al., 2000), it appears that even by the age of 3 years children were reactive to the intake-enhancing effects of the mere presence of large portion sizes. The children who ate more when served large portions also tended to eat more in the absence of hunger, probably because their food consumption was primarily triggered by environmental cues such as the mere presence and availability of food and perhaps also by being rewarded for cleaning their plates. However, when children were allowed to self-select their own portion of food, they consumed 25% less food than when large portions were preserved on their plates.

The trend toward "supersizing" of food portions is being driven by a number of factors. On the one hand, large portion sizes are relatively cheap for the food manufacturing, food processing, and food service industries to provide. From the consumer's point of view, larger portion sizes represent good value for money. People also tend to eat in units (Rolls, 2003) and, as a result, eat what is served, even though it may be wholly incompatible with their true energy needs.

Although it cannot be definitively stated that the concurrent trends in fast food consumption, snacking, and increasing portion size and obesity are causally related, it is inconceivable that they are not linked, even if these links remain speculative. However, it is highly unlikely that the shifts in EI as a result of these eating behaviors have been solely responsible for the present obesity trends.

CONCLUSIONS

Until recently, pediatric obesity was not considered to be a public health problem and did not feature highly on the public agenda. Several factors have now combined to make the issue a public health priority, which will require a concerted societal response if the problem is to be attenuated and eventually reversed. These factors include the escalating prevalence rates of childhood obesity, the recognition that obese children tend to become obese adults, and the deleterious short- and long-term social, economic, and health consequences of childhood obesity. Consequently, there is now consensus that the need to improve children's dietary and physical activity patterns must be tackled more aggressively than hitherto, because failure to do so will have catastrophic economic and health implications in the future.

Currently, childhood obesity is one of the hottest nutrition topics in the media. However, although this has heightened public awareness of the problem, it has also fueled an unhelpful tendency to target culpability at specific food(s). It needs to be emphasized that obesity results more from overall eating patterns rather than from consumption of specific food stuffs. This in turn suggests that it is counterproductive to target specific foods for dietary advice as this only serves to dichotomize foods into "good" and "bad." Given that the latter are usually energy dense foods high in fat and/or sugar, it has been demonstrated that restricting access to such foods is likely to increase children's preferences for and consumption of them when restriction is removed (Fisher and Birch, 1999). Paradoxically, although not intended, parents may inadvertently promote children's liking for energy dense foods through overly restrictive child-feeding practices.

For better or worse, all the risk factors for the development of childhood obesity have their initial beginnings in the family of origin. Birch and Fisher (1998) have provided extensive evidence that childhood food preferences and eating patterns may be permanently modeled by parental eating patterns and food beliefs. Insights gained from this work could be used to inform the development of interventions designed to improve child-feeding practices, but, unfortunately, the current food environment often forces parents into a no-win situation with respect to child-feeding practices.

The locus of responsibility for childhood obesity does not rest solely with individuals. Regrettably, most dietary (and physical activity) prescriptions for the treatment and prevention of childhood obesity have had spectacularly unsuccessful results. This is likely to remain the case until it is fully appreciated and accepted that there are multiple and interrelated determinants of children's eating and physical activity behaviors, including biological, parental, and societal influences. Without radical measures to detoxify the current obesogenic environment, rates of childhood obesity will never be arrested, let alone reversed.

REFERENCES

American Academy of Pediatrics, Soft drinks in schools. Policy Statement, *Pediatrics,* 113, 152–154, 2004.

American Institute for Cancer Research, *New Survey Shows Americans Ignore Importance of Portion Size in Managing Weight,* available at http://www.aicr.org/obesity, March 2000.

Astrup, A. and Raben, A., Carbohydrate and obesity, *Int. J. Obes.*, 19, S27–S37, 1995.

Bellisle, F., Rolland-Cachera, M.F., Deheeger, M., and Guilloud-Bataille, M., Obesity and food intake in children: evidence for a role of metabolic and/or behavioral daily rhythms, *Appetite*, 11, 111–118, 1988.

Birch, L.L., Children's preferences for high-fat foods, *Nutr. Rev.*, 50, 249–255, 1992.

Birch, L.L. and Deysher, M., Conditioned and unconditioned caloric compensation: evidence for self-regulation of food intake by young children, *Learning Motiv.*, 16, 341–355, 1985.

Birch, L.L., Johnson, S.L., Andresen, G., Peters, J.C. and Schulte, M.C., The variability of young children's energy intake, *N. Engl. J. Med.*, 324, 232–235, 1991.

Birch, L.L., Johnson, S.L., Jones, M.B. and Peters, J.C., Effects of a nonenergy fat substitute on children's energy and macronutrient intake, *Am. J. Clin. Nutr.*, 58, 326–333, 1993.

Birch, L.L. and Fisher, J.O., Development of eating behaviors among children and adolescents, *Pediatrics*, 101, 539–549, 1998.

Birch, L.L. and Fisher, J.O., Mothers' child-feeding practices influence daughters' eating and weight, *Am. J. Clin. Nutr.*, 71, 1054–1061, 2000.

Blundell, J.E. and Stubbs, R.J., High and low carbohydrate and fat intakes: limits imposed by appetite and palatability and their implications for energy balance, *Eur. J. Clin. Nutr.*, 53, S148–S165, 1999.

Cavadini, C., Siega-Riz, A.M. and Popkin, B.M., US adolescent food intake trends from 1965 to 1996, *Arch. Dis. Child.*, 83, 18–24, 2000.

Davies, P.S.W., Diet composition and body mass index in pre-school children, *Eur. J. Clin. Nutr.*, 51, 443–448, 1997.

Davis, C.M., Self-selection of diet by newly weaned infants, *Am. J. Dis. Child.*, 36, 651–679, 1928.

DiMeglio, D.P. and Mattes, R.D., Liquid versus solid carbohydrate: effects on food intake and body weight, *Int. J. Obes.*, 24, 794–800, 2000.

Drewnowski, A., Macronutrient substitutes and weight-reduction practices of obese, dieting and eating-disordered women, *Ann. N.Y. Acad. Sci.*, 819, 132–141, 1997.

Drummond, S., Crombie, N. and Kirk, T.A., A critique of the effects of snacking on body weight status, *Eur. J. Clin. Nutr.*, 50, 779–783, 1996.

Edelman, B., Engell, D., Bronstein, P. and Hirsch, E., Environmental effects on the intake of overweight and normal-weight men, *Appetite*, 7, 71–83, 1986.

Fisher, J.O. and Birch, L.L., Fat preferences and fat consumption of 3- to 5-year-old children are related to parental adiposity, *J. Am. Diet. Assn.*, 95, 759–764, 1995.

Fisher, J.O. and Birch, L.L., Restricting access to palatable foods affects children's behavioral response, food selection, and intake, *Am. J. Clin. Nutr.*, 69, 1264–1272, 1999.

Fisher, J.O., Rolls, B.J., and Birch, L.L., Children's bite size and intake of an entrée are greater with large portions than with age-appropriate or self-selected portions, *Am. J. Clin. Nutr.*, 77, 1164–1170, 2003.

French, S.A., Story, M., Neumark-Sztainer, D., Fulkerson, J.A., and Hannan, P., Fast food restaurant use among adolescents: associations with nutrient intake, food choices and behavioral and psychosocial variables, *Int. J. Obes.*, 25, 1823–1833, 2001.

Gatenby, S.J., Eating frequency: methodological and dietary aspects, *Br. J. Nutr.*, 77 (Suppl. 1), S7–S20, 1997.

Gazzaniga, J.M. and Burns, T.L., Relationship between diet composition and body fatness, with adjustment for resting energy expenditure and physical activity in preadolescent children, *Am. J. Clin. Nutr.*, 58, 21–28, 1993.

Gibson, S.A., Are high-fat, high-sugar foods and diets conducive to obesity? *Int. J. Food Sci. Nutr.*, 47, 405–415, 1996.

Gibson, S.A., Associations between energy density and macronutrient composition in the diets of pre-school children: sugars vs. starch, *Int. J. Obes.*, 24, 633–638, 2000.

Guillaume, M., Lapidus, L., and Lambert, A., Obesity and nutrition in children. The Belgian Luxembourg Child Study IV, *Eur. J. Clin. Nutr.*, 52, 323–328, 1998.

Harnack, L., Stang, J., and Story, M., Soft drink consumption among US children and adolescents: nutritional consequences, *J. Am. Diet. Assn.*, 99, 436–441, 1999.

Hill, J.O. and Prentice, A.M., Sugar and body weight regulation, *Am. J. Clin. Nutr.*, 62, 264S– 274S, 1995.

Hill, A.J., Oliver, S., and Rogers, P.J., Eating in the adult world: the rise of dieting in childhood and adolescence, *Br. J. Clin. Psychol.*, 31, 95–105, 1992.

Jahns, L., Siega-Riz, A.M. and Popkin, B.M., The increasing prevalence of snacking among US children from 1977–1996, *J. Pediatr.*, 138, 493–498, 2001.

Johnson, S.L., McPhee, L., and Birch, L.L., Conditioned preferences: young children prefer flavors associated with high dietary fat, *Physiol. Behav.*, 50, 1245–1251, 1991.

Kant, A.K., Schatzkin, A., Graubard, B.I., and Ballard-Barbash, R., Frequency of eating occasions and weight change in NHANES. I. Epidemiologic follow-up study, *Int. J. Obes.*, 19, 468–474, 1995.

Klesges, R.C., Klesges, L.M., Eck, L.H., and Shelton, M.L., A longitudinal analysis of accelerated weight gain in preschool children, *Pediatrics*, 95, 126–130, 1995.

Ludwig, D.S., Majzoub, J.A., Al-Zahrani, A., Dallal, G.E., Blanco, I., and Roberts, S.B., High glycemic index foods, overeating, and obesity, *Pediatrics*, 103, e26, available at http://www.pediatrics.org/cgi/content/full/103/3/e26 (accessed 7 January 2004), 1999.

Ludwig, D.S., Dietary glycemic index and obesity, *J. Nutr.*, 130, 280S–283S, 2000.

Ludwig, D.S., Peterson, K.E., and Gortmaker, S.L., Relation between consumption of sugar-sweetened drinks and childhood obesity: a prospective, observational analysis, *Lancet*, 357, 505–508, 2001.

Maffeis, C., Talamini, G., and Tato, L., Influence of diet, physical activity and parents' obesity on children's adiposity: a four-year longitudinal study, *Int. J. Obes.*, 22, 758–764, 1998.

Magarey, A.M., Daniels, L.A., Boulton, T.J.C., and Cockington, R.A., Does fat intake predict adiposity in healthy children and adolescents aged 2–15 y? A longitudinal analysis, *Eur. J. Clin. Nutr.*, 55, 471–481, 2001.

Mattes, R.D., Dietary compensation by humans for supplemental energy provided as ethanol or carbohydrate in fluids, *Physiol. Behav.*, 59, 179–187, 1996.

McConahy, K.L., Smiciklas-Wright, H., Birch, L.L., Mitchell, D.C., and Picciano, M.F., Food portions are positively related to energy intake and body weight in early childhood, *J. Pediatr.*, 140, 340–347, 2002.

McGloin, A.F., Livingstone, M.B.E., Greene, L.C., Webb, S.E., Gibson, J.M.A., Jebb, S.A., Cole, T.J., Coward, W.A., Wright, A., and Prentice, A.M., Energy and fat intake in obese and lean children at varying risk of obesity, *Int. J. Obes.*, 26, 200–207, 2002.

Ministry of Agriculture, Fisheries and Food/Department of Health., *National Diet and Nutrition Survey: Young People Aged 4 to 18 Years*, The Stationery Office, London, 2000.

Mrdjenovic, G. and Levitsky, D.A., Nutritional and energetic consequences of sweetened drink consumption in 6- to 13-year-old children, *J. Pediatr.*, 142, 604–610, 2003.

Nguyen, V.T., Larson, D.E., Johnson, R.K., and Goran, M.I., Fat intake and adiposity in children of lean and obese parents, *Am. J. Clin. Nutr.*, 63, 507–513, 1996.

Nielsen, S.A., Siega-Riz, A.M., and Popkin, B.M., Trends in food locations and sources among adolescents and young adults, *Prev. Med.*, 35, 107–113, 2002.

Obarzanek, E., Schreiber, G.B., Crawford, P.B., Goldman, S.R., Barrier, P.M., Frederick, M.M., and Lakatos, E., Energy intake and physical activity in relation to indexes of body fat: the National Heart, Lung, and Blood Institute Growth and Health Study, *Am. J. Clin. Nutr.*, 60, 15–22, 1994.

Ortega, R.M., Requejo, A.M., Andres, P., Lopez-Sobaler, A.M., Redondo, R., and Gonzalez-Fernandez, M., Relationship between diet composition and body mass index in a group of Spanish adolescents, *Br. J. Nutr.*, 74, 765–773, 1995.

Pawlak, D.B., Ebbeling, C.B., and Ludwig, D.S., Should obese patients be counselled to follow a low glycaemic index diet? Yes, *Obes. Rev.*, 3, 235–243, 2002.

Prentice, A.M. and Jebb, S.A., Fast foods, energy density and obesity: a possible mechanistic link, *Obes. Rev.*, 4, 187–194, 2003.

Raben, A., Macdonald, I., and Astrup, A., Replacement of dietary fat by sucrose or starch: effect of 14 d ad libitum energy intake, energy expenditure and body weight in formerly obese and never-obese subjects, *Int. J. Obes.*, 21, 846–859, 1997.

Raben, A., Should obese patients be counselled to follow a low glycaemic index diet? No, *Obes. Rev.*, 3, 245–256, 2002.

Remer, T., Dimitriou, T., and Kersting, M., Does fat intake explain fatness in healthy children? *Eur. J. Clin. Nutr.*, 56, 1046–1047, 2002.

Ricketts, C.D., Fat preferences, dietary fat intake and body composition in children, *Eur. J. Clin. Nutr.*, 51, 778–781, 1997.

Roberts, S.B., High-glycemic index foods, hunger, and obesity: Is there a connection? *Nutr. Rev.*, 58, 163–169, 2000.

Robertson, S.M., Cullen, K.W., Baranowski, J., Baranowski, T., Hu, S., and de Moor, C., Factors related to adiposity among children aged 3 to 7 years, *J. Am. Diet. Assn.*, 99, 938–943, 1999.

Rolland-Cachera, M.F., Deheeger, M., Akrout, M., and Bellisle, F., Influence of macronutrients on adiposity development: a follow up study of nutrition and growth from 10 months to 8 years of age, *Int. J. Obes.*, 19, 573–578, 1995.

Rolls, B.J., Bell, E.A., Castellanos, V.H., Chow, M., Pelkman, C.L., and Thorwart, M.L., Energy density but not fat content of foods affected energy intake in lean and obese women, *Am. J. Clin. Nutr.*, 69, 863–871, 1999.

Rolls, B.J., Engell, D., and Birch, L.L., Serving portion size influences 5-year-old but not 3-year-old children's food intakes, *J. Am. Diet. Assn.*, 100, 232–234, 2000.

Rolls, B.J., Morris, E.L., and Roe, L.S., Portion size of food affects energy intake in normal-weight and overweight men and women, *Am. J. Clin. Nutr.*, 76, 1207–1213, 2002.

Rolls, B.J., The Supersizing of America, *Nutr. Today*, 38, 42–53, 2003.

Saris, W.H.M., Astrup, A., Prentice, A.M., Zunft, H.J., Formiguera, X., Verboeket-van de Venne, W.P., Raben, A., Poppitt, S.D., Seppelt, B., Johnston, S., Vasilaras, T.H., and Keogh, G.F., Randomized controlled trial of changes in dietary carbohydrate/fat ratio and simple vs. complex carbohydrates on body weight and blood lipids: the CARMEN study, *Int. J. Obes.*, 24, 1310–1318, 2000.

Siega-Riz, A.M., Carson, T., and Popkin, B., Three squares or mostly snacks—what do teens really eat? A sociodemographic study of meal patterns, *J. Adolesc. Health*, 22, 29–36, 1998.

Spieth, L.E., Harnish, J.D., Lenders, C.M., Raezer, L.B., Pereira, M.A., Hangen, J., and Ludwig, D.S., A low-glycemic index diet in the treatment of pediatric obesity, *Arch. Pediatr. Adolesc. Med.,* 154, 947–951, 2000.

Story, M. and Brown, J.E., Do young children instinctively know what to eat? The studies of Clara Davis revisited, *N. Engl. J. Med.* 316, 103–105, 1987.

Stubbs, R.J., Harbron, C.G. and Prentice, A.M., Covert manipulation of the dietary fat to carbohydrate ratio of isoenergetically dense diets: effect on food intake in men feeding ad libitum, *Int. J. Obes.,* 20, 651–660, 1996.

Stubbs, R.J., Johnstone, A.M., Harbron, C.G., and Reid, C., Covert manipulation of energy density of high carbohydrate diets in "pseudo free-living" humans, *Int. J. Obes.,* 22, 885–892, 1998.

Stubbs, R.J., Mazlan, N., and Whybrow, S., Carbohydrates, appetite and feeding behavior in humans, *J. Nutr.,* 131, 2775S–2781S, 2001.

Summerbell, C.D., Moody, R.C., Shanks, J., Stock, M.J., and Geissler, C., Relationship between feeding pattern and body mass index in 220 free-living people in four age groups, *Eur. J. Clin. Nutr.,* 50, 513–519, 1996.

Troiano, R.P., Briefel, R.R., Carroll, M.D., and Bialostosky, K., Energy and fat intakes of children and adolescents in the United States: data from the National Health and Nutrition Examination Surveys, *Am. J. Clin. Nutr.,* 72, 1343S–1353S, 2000.

Tucker, L.A., Seljaas, G.T., and Hager, R.L., Body fat percentage of children varies according to their diet composition, *J. Am. Diet. Assn.,* 97, 981–986, 1997.

Wardle, J., Guthrie, C., Sanderson, S., Birch, L., and Plomin, R., Food and activity preferences in children of lean and obese parents, *Int. J. Obes.,* 25, 971–977, 2001a.

Wardle, J., Sanderson, S., Gibson, E.L., and Rapoport, L., Factor-analytic structure of food preferences in four-year-old children in the UK, *Appetite,* 37, 217–223, 2001b.

Warren, J.M., Henry, J.K., and Simonite, V., Low glycemic index breakfasts and reduced food intake in preadolescent children, *Pediatrics,* 112, e414–e419, available at http://www.pediatrics.org/cgi/content/full/112/5/e414 (accessed 5 January 2004), 2003.

World Health Organization/Food and Agriculture Organization Expert Consultation Diet, Nutrition and the Prevention of Chronic Disease, *Diet, Nutrition and the Prevention of Chronic Disease: Report of a Joint WHO/FAO/ Expert Consultation, Geneva, 28 January–1 February 2002,* World Health Organization, Geneva, Switzerland, WHO Technical Report Series No. 19, 2003.

Young, L.R. and Nestle, M., The contribution of expanding portion sizes to the US obesity epidemic, *Am. J. Public Health,* 92, 246–249, 2002.

9

CHILDHOOD OBESITY,
PHYSICAL ACTIVITY, AND
THE ENVIRONMENT

Ashley R. Cooper and Angie Page

CONTENTS

INTRODUCTION

Obesity is increasing worldwide in both adults and children, to the extent that the World Health Organization recently declared that obesity was a pandemic that constituted one of the leading future threats to public health (WHO, 2002). In the U.S., the proportion of children and adolescents who are overweight has tripled in the past three decades (Ogden et al., 2002), and similar changes are being seen in the U.K. Between 1984 and 1994, the prevalence of obesity in children in the U.K. increased two- to threefold (Hughes et al., 1997), and the National Diet and Nutrition Survey (2000) found that for respondents ages 8 to 18 years 15.4% were overweight and 4% were obese (Jebb et al., 2003). Although childhood obesity is less prevalent than adult obesity, it nonetheless gives cause for concern. It is associated with a less favorable cardiovascular risk profile, including hypertension, hyperlipidemia, and glucose intolerance (Must and Strauss, 1999), a low health-related quality of life (Schwimmer et al., 2003) and predisposes children to many of the medical complications of obesity found in adults that may lead to serious chronic disease (Chu et al., 1998; Sorof et al., 2002). Additionally, fatter children are more likely to become fatter and less healthy adults, further increasing adult obesity levels in the future (Clarke and Lauer, 1993).

One of the major causes of this increase in obesity is believed to be an environment that encourages sedentary behavior in combination with overconsumption of energy-dense foods. The relative contribution of these factors is not known, partly because accurate measurements of food and drink consumption or physical activity are extremely difficult, although both are likely to be significantly involved. Given the difficulty of measurement, it is maybe not surprising that there are no longitudinal data in the U.K. describing population changes in physical activity. However, transportation surveys show that children are walking and cycling less, and surrogate measures of inactivity — car, TV, and computer game ownership — are all increasing (DETR, 2000; Prentice and Jebb, 1995). In particular, increased use of the car for even the shortest journey gives cause for concern.

In this chapter, we describe advances in the measurement of physical activity in children that allow us to investigate the patterns of children's activities and the possible associations of these with obesity. Because many children are sufficiently active for health, such profiling is valuable in identifying children who may benefit from intervention to increase activity levels; knowledge of *how* active children are and *when* and *where* this activity occurs will also allow us to develop targeted interventions for such individuals. In addition, such measurements allow broader interventions to be assessed. Public health agencies have begun to implement schemes to try and reverse the trend of diminishing activity, and the promotion of active travel to school is one such initiative (DETR, 1999). Surprisingly, this intervention has been implemented with little evidence of efficacy. We will describe recent data using activity profiling to investigate the association between active travel to school and overall physical activity.

OBJECTIVE MEASUREMENT OF CHILDREN'S ACTIVITY

Despite the high level of concern about the adverse consequences of low physical activity in children, the levels and patterns of children's physical activity and how these influence children's present and future health are remarkably underresearched. Indeed, for the majority of children, there may be no cause for concern. In the latest national survey in England, the large majority of children appeared to be meeting current recommendations of 60 minutes of moderate-intensity physical activity each day (Sproston and Primatesta, 2003). According to these standards, over two thirds of children and adolescent boys are sufficiently active; adolescent girls are of more concern, with 50% or less meeting guidelines. This survey, which relied on self-report data, may be susceptible to bias due to socially desirable reporting; however, results are similar to findings from smaller studies that used objective assessments of physical activity in England (Cooper et al., 2003), the U.S. (Pate et al., 2002), and other European countries (Riddoch et al., 2004). It thus appears that it is only a minority (with the exception of teenage girls) who need attention. However, it should also be noted that the current recommendations are based on weak evidence and may be set too low (Twisk, 2001).

Although health status during youth is a predictor of health status in adulthood (Twisk et al., 1997), there are relatively little data linking adolescent health status to physical activity. The strongest associations are found between activity and body fatness (Bar-Or and Baranowski, 1994) and activity and cardiopulmonary fitness (Morrow and Freedson, 1994). Few young people demonstrate overt ill health due to inactivity, although an adverse metabolic profile may be beginning to be established. For example, in otherwise healthy Danish children and adolescents, low physical fitness is associated with clustering of adverse physiological indicators (Andersen et al., 2003), and in obese adolescent boys the presence of multiple cardiovascular risk factors has been associated with low physical performance (Torok et al., 2001). However, it is clear that most young people are not insufficiently active to be at immediate risk of poor health, and thus we need to be able to identify and prioritize those who would benefit from intervention to increase physical activity. Individual profiling of physical activity will allow us to do this; in addition, this will enable targeting of such interventions to periods when other more active children carry out activities that enable them to reach and exceed recommended guidelines.

PROFILING OF CHILDREN'S PHYSICAL ACTIVITY

Children accumulate physical activity in varied ways, including play, organized games and sports, and active transport. Playing includes running, jumping, throwing, catching, and chasing, all activities that are highly intermittent and spontaneous, sporadic both in time and intensity (Bailey et al., 1995). As children get older, the balance between these different activities changes, with play contributing a smaller proportion and sedentary pursuits increasingly taking up their time. Measurement of such a diverse range of activity is challenging, and until recently this dilemma has limited our understanding of the relationship between physical activity and children's health. It is clearly naïve to see physical activity as a single entity, especially where we are investigating potential solutions to inactivity, and profiles of physical activity reflecting different elements and settings may offer a more helpful picture.

The majority of investigations of children's activity utilize self-reporting measurements such as questionnaires, which provide only crude estimates of levels of activity. To provide more detailed measurements of patterns of activity intensity, heart rate monitoring has been widely used to investigate physical activity during the school day or to compare school and weekend days (Armstrong, 1998). Although heart rate monitoring is a useful technique for assessing moderate to vigorous activity, it has limited value in providing information on general activity levels, and the results of such studies are equivocal. More recently, accelerometers able to record the activity level of children each minute of the day for a sustained period of several weeks have been developed. These instruments allow us to produce a profile of individual or group activity levels and patterns with high detail and are increasingly being used to help us understand when children are most or least physically active, thus aiding the design of effective interventions.

Physical Activity Within the School Environment

The school day is seen as the focus for understanding children's activities, since during the week a large part of their day is spent at school. Children spend the majority of their time sitting at school, with physical activity only occurring mainly at break times and during physical education classes. However, studies utilizing direct observation have shown that, although relatively brief, these periods of physical activity may be important for children's health, since it is during these times that children carry out their highest daily levels of activity with regard to the intensity of activity (Sleap and Warburton, 1996). Similarly, our studies using the CSA accelerometer in over 200 English primary school children between 10 and 12 years old found that the lunch break provides the highest amount of moderate to vigorous physical activity (MVPA) achieved during the school day (Figure 9.1) or indeed at any time during the week. This spontaneous activity during morning and lunch breaks contributed approximately 22% (39 minutes) and 19% (27 minutes) of total daily MVPA for boys and girls, respectively. In older children (12–15 years), overall levels of physical activity were lower, but the contribution of break times to total MVPA was similar.

These active breaks may have further, indirect benefits. In a study of seventy-six 9-year-old children, objective measurements using accelerometers

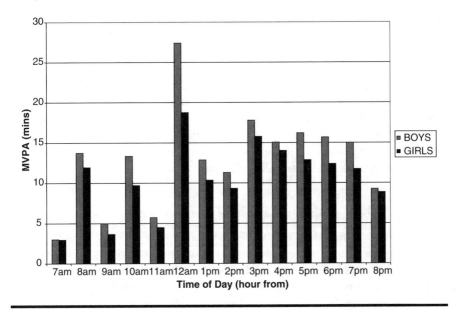

Figure 9.1   School Day Pattern of Moderate to Vigorous Physical Activity (MVPA) in English Primary School Children.

were used to assess the influence of experimental manipulations of the children's day. On days where children worked at computers during breaks and had no physical education classes, after school physical activity was significantly lower than days where children participated in outdoor breaks and physical education (Dale et al., 2000). Thus, it is important for children to be encouraged to participate in physical activity during the school day. From a public health perspective, it is of serious concern that lack of playground supervision and concerns over bullying or personal injury are encouraging schools to reduce the amount of time for play during breaks.

Physical education classes have traditionally been viewed as a primary source of children's physical activity. However, the contribution of physical education to overall physical activity is a matter of debate, as it takes up less than 1% of a child's waking time. Even during classes, children may not be very active—studies of English children (ages 5 to 11 years) using direct observations of activities during physical education classes have shown that nearly one quarter of lesson time is taken up with physically passive activities, and less than one third of children are engaged in a 5-minute sustained period of MVPA (Sleap and Warburton, 1996). To investigate the influence of differing amounts of timetabled physical education on overall physical activity, Mallam et al. (2003) used accelerometers to compare over 200 English primary school children from three schools, with one having excellent sport facilities and 9 hours per week of scheduled physical educations and sports and the other two providing approximately 2 hours of physical education per week. Surprisingly, the total amount of physical activity carried out by children from all three schools was similar, with those from the schools providing less physical education having significantly more out of school activity. The role of physical education in directly providing a significant contribution to children's daily physical activity is thus questionable; perhaps the limited time for physical education should be spent using educational and psychological strategies to influence children's use of time beyond the confines of the lesson. Such approaches may ultimately increase child/youth activity levels and further translate through to adulthood.

Physical Activity Outside the School Environment

Although the school has been the focus of many studies, activities that occur outside school hours are a major contributor to children's overall physical activity level. Our profiling data in primary school children identified that, on school days, both boys and girls recorded nearly one half of their total daily physical activity between 3:00 and 8:00 P.M. (Figure 9.1). Weekend activity levels were very different from this, with the children being significantly less active during the weekend than during school days. The data also demonstrated a markedly different pattern from that seen during the school day, with no peaks of activity but rather a more general inverse U-shaped curve, with levels highest in midafternoon. Similar data have been reported in U.S. children and adolescents (Trost et al., 2000), indicating that weekend leisure time may be an appropriate target for physical activity intervention programs.

HOW PHYSICALLY ACTIVE ARE OBESE CHILDREN?

Although the role of physical inactivity as a behavioral risk factor for the development of childhood obesity has yet to be unequivocally determined, much research has investigated differences in physical activities between those children who are already obese and their nonobese peers. Most, although not all, cross-sectional survey data have suggested that obese or overweight children spend less time in MVPA than their nonobese counterparts, and this observation has been supported by smaller studies that used objective physical activity measurements. Obese adolescents (Ekelund et al., 2002), school-aged children (Trost et al., 2001), and overweight preschool children (Trost et al., 2003) have all been shown to be less physically active than their nonobese or nonoverweight peers, consistent with the idea that physical inactivity is an important contributing factor in the maintenance of obesity. Such studies have however reported mean differences between groups of individuals and have not considered when such differences may occur or indeed the differences in physical activities between individuals.

The major contribution made to overall physical activity outside school suggests that differences in physical activity associated with health outcomes may become apparent in this period. To investigate this, we measured the daily physical activity patterns of obese and nonobese children using accelerometers. Children were recruited from schools in Bristol and from the childhood obesity clinic, Bristol Royal Hospital for Children, Bristol, and wore an accelerometer while carrying out their normal daily activities for 7 days; 249 children (mean age 10.5 ± 0.8 years) took part in the study, with 23.3% above the 95th percentile for body-mass index (BMI).

A gender difference was seen in overall activity levels, with the girls being approximately 20% less active than the boys. In both genders, the obese children were significantly less physically active overall than their nonobese counterparts and spent less time in physical activity of moderate or greater intensity. These differences were seen for almost every hour of the week and weekend day, but the major differences in activity between obese and nonobese children occurred outside of school time, between 3:00 and 8:00 P.M. (Figure 9.2, A and B), and during weekends (Figure 9.3, A and B).

Hourly patterns of activity indicate that obese children tend to be less active at times when activity is more likely to be determined by free choice, particularly outside of school. Such patterns of physical activity may have contributed to and are likely to sustain their obesity. Activity profiling thus allows us to identify periods when intervention to increase activity may be most appropriate. A further question is whether all obese children are physically inactive and are thus appropriate targets for intervention. Fortunately, accelerometry can also help us to investigate this issue.

Figure 9.4, A and B, shows the range of activity above and below mean values for the boys and girls in the previous study related to BMI centile. It is clear that most, but not all, children above the 95th centile are less active than the population mean; however, there are also many children of normal BMI who have low levels of physical activity. It is thus possible to be an obese, yet active, child or a normal weight but relatively sedentary child. At present, we do not know the future consequences of physical activity levels for either of these groups.

It is thus possible and valuable to use accelerometry to produce activity profiles for children. These enable us to identify sedentary groups of individuals and to describe when and where interventions to increase physical activity might be implemented. The promotion of active transport to school is perhaps one of the largest natural interventions, and we have begun to use this technique to investigate the influence that this may have on physical activity levels and health.

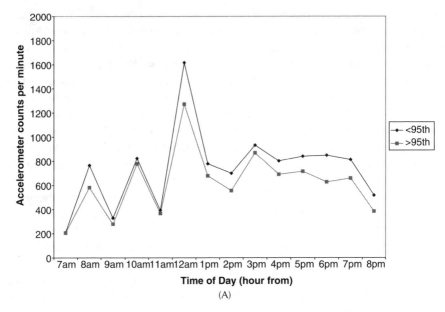

(A)

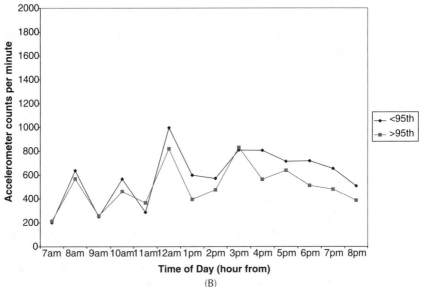

(B)

Figure 9.2 A: School Day Pattern of Physical Activity in Obese (>95th Centile for BMI) vs. Nonobese (<95th Centile for BMI) Primary School-Aged Boys. B: School Day Pattern of Physical Activity in Obese (>95th Centile for BMI) vs. Nonobese (<95th Centile for BMI) Primary School-Aged Girls.

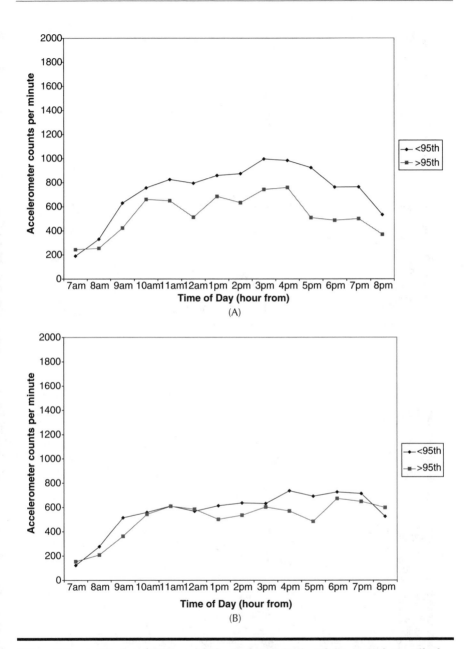

Figure 9.3 A: Weekend Pattern of Physical Activity in Obese (>95th Centile for BMI) vs. Nonobese (<95th Centile for BMI) Primary School-Aged Boys. B: Weekend Pattern of Physical Activity in Obese (>95th Centile for BMI) vs. Nonobese (<95th Centile for BMI) Primary School-Aged Girls.

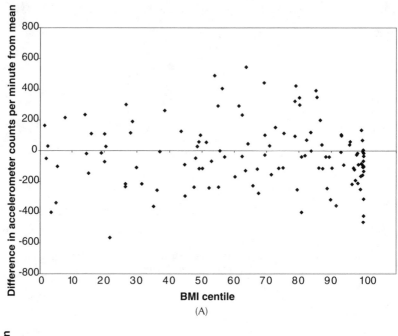

(A)

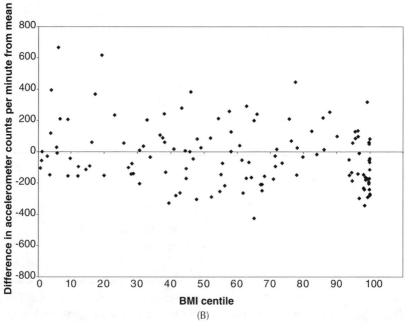

(B)

Figure 9.4   A: Distribution of Physical Activity Level by BMI Centile for Boys (10.5 Years). B: Distribution of Physical Activity Level by BMI Centile for Girls (10.5 Years).

ACTIVE TRANSPORT TO AND FROM SCHOOL

Children are naturally spontaneously active, but parental concerns about the safety of the environment such as heavy traffic and fear of abduction or bullying, busier lifestyles, more single parent families, and wider school choices meaning longer distances to school have led to more parents opting to use the car as a form of transport for even short journeys. Consequently, the proportion of children walking and cycling to school has fallen markedly in the past 15 years. The National Travel Survey (DETR, 2000) found that in 1985/6 nearly 60% of school journeys made by children (5–16 years) were on foot and that only 16% were by car; just over a decade later, however, the proportion of children walking had fallen to 48% and the proportion driven to school had nearly doubled to 30%. Cycling to school is now an unusual activity — during the same period, cycling to school fell from 3.5% to less than 1%. These changes suggest that not only the amount of activity that children get has been reduced but a dependency on the car for transport has developed, a behavior that continues into adulthood.

In an attempt to address this issue, the journey to school has been recognized by the U.K. government as an important opportunity for establishing daily physical activity for children, and many schemes have been developed to promote active travel to school (DETR, 1999; Rowland et al., 2003). Surprisingly, until recently, no studies had investigated whether active travel to school could influence children's overall physical activity. Although intuitively it seems clear that active transport will increase physical activity, it is possible that children who walk to school may be no more active overall than those who travel by car. The car users may have more playtime due to shorter journeys or the children who walk may be tired and compensate with less activity during the remainder of the day. Alternatively, the children who choose to walk may be more active than expected through greater social contact or greater independent mobility or given greater license to be active from parents.

ACTIVE TRAVEL AND PHYSICAL ACTIVITY PATTERNS

We have recently investigated this issue by measuring the activity profiles of 114 children from five urban primary schools using an accelerometer and investigating differences associated with mode of transport to school (Cooper et al., 2003). The only difference between travel groups in girls was when going to school, with those on foot being more active during that time. However, in boys, substantial differences in the amount of activity between those who walked to school and those who were driven were found, more than could be accounted for by the journey alone. To find out when this difference occurred, we plotted the amount of physical

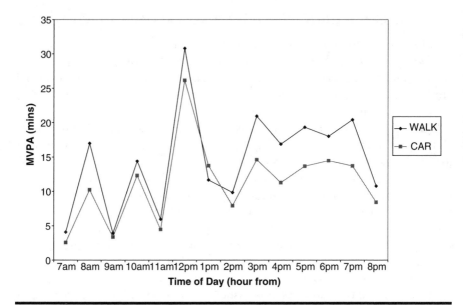

Figure 9.5   Level of Moderate to Vigorous Physical Activity (MVPA) of Boys Who Walk to School Compared With Those Traveling to School by Car.

activity that the groups of boys did for each hour of the day (Figure 9.5). The boys who walked to school were more active between 8:00 and 9:00 A.M., showing the effect of the journey, but there were no significant differences between the travel groups while in school. However, after school and during the evening, boys who walked to school were consistently more active than those who traveled by car.

Using physical activity diaries to investigate the source of these differences, we found that boys who walked to school had higher levels of play, particularly in games such as football, than those traveling by car (Figure 9.6). These data suggest that active transport may provide the opportunity for children to utilize local resources for play and may act as a catalyst for increasing informal and spontaneous aspects of physical activity.

CHILDREN'S PHYSICAL ACTIVITY AND HEALTH

Just how great are these differences between the children who travel in different ways? U.K. national guidelines encourage children to accumulate at least 60 minutes per day of moderate physical activity, such as brisk walking, swimming, or cycling (Biddle et al., 1998). In this study, the journey to and from school contributed 10 to 15 minutes per day of moderate activity for those who walked. In addition to this, boys using active travel carried out a further 30 minutes each day of moderate-intensity

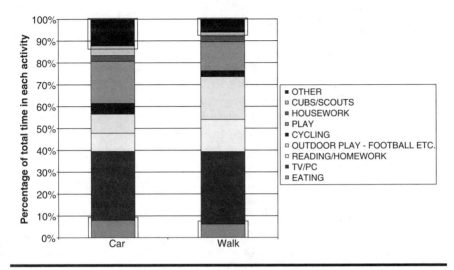

Figure 9.6   Time Spent in Different Activities Between 3:00 and 8:00 P.M. in Boys, Classified by Mode of Travel to School.

activity compared with car users. Clearly, active travel to school has the potential to substantially contribute to the achievement of physical activity guidelines, but the direct health benefits remain to be demonstrated. We estimate that the energy expended in extra physical activity by the boys is equivalent to approximately 2 kg weight per year, with all other things being equal. Over the years, this difference could significantly contribute to the development of obesity, and even at this young age the boys who used active transport were slightly leaner.

CONCLUSION

Activity profiling is beginning to allow us to understand the physical activity patterns of children and young people in more detail. Data suggesting that active travel may encourage higher physical activity are exciting, and we are planning further studies to describe the after school and evening activity of children and the social and environmental context in which they do these activities. Further data to support the benefit of physically active commuting in children are also beginning to emerge. In Denmark, boys who walk or cycle to school are more active during supervised play than those who travel by car, and Filipino adolescents who walk to school have been shown to have greater daily energy expenditure than their peers who travel by car. Norwegian studies have found that children who use active transport to school have greater physical fitness than those who use the bus. These data strongly suggest that we must provide opportunities for children to be active around the school day, and development of an infrastructure that encourages walking and cycling to school must be seen as a key component of that.

Unfortunately, such ideas are not central to policy. One contributor to the increase in car travel is thought to be increased separation between home and school due to greater parental choice of school. This is particularly marked for secondary schools where the mean distance traveled between home and school increased from 2.1 miles in 1985/6 to 3.5 miles in 1997/9. A journey of 2 miles will take 30 to 40 minutes to walk, but a near doubling of this distance—walking to school and back again afterward—makes this an unrealistic prospect for many. Cycling to school for secondary children would thus seem to be a better option than walking given the distances involved. However, the conditions around schools for cycling have worsened due to increased vehicle traffic, with almost 20% of cars in urban areas doing the school run during the morning peak. This volume of traffic, without dedicated cycle ways or effective traffic management schemes that give pedestrians/cyclists priority, makes cycling unpleasant and potentially dangerous.

Integration of safe routes to school such as cycle paths protected from the traffic and walking trails away from roads would be a policy and environmental intervention that could change passive transport to an active journey for large groups of children. Such a change may facilitate increased physical activity in children and perhaps contribute toward prevention of childhood obesity.

REFERENCES

Andersen, L.B., Wedderkopp, N., Hansen, H.S., Cooper, A.R., and Froberg, K., Biological cardiovascular risk factors cluster in Danish children and adolescents: the European Youth Heart Study, *Prev. Med.*, 37, 363–367, 2003.

Armstrong, N., Young people's physical activity patterns as assessed by heart rate monitoring, *J. Sports Sci.*, 16, S9–S16, 1998.

Bailey, R.C., Olson, J., Pepper, S.L., Porszaz, J., Barstow, T.L., and Cooper, D.M., The level and tempo of children's physical activities: an observation study, *Med. Sci. Sports Exerc.*, 27, 1033–1041, 1995.

Bar-Or, O. and Baranowski, T., Physical activity, adiposity, and obesity among adolescents. *Pediatric Exercise Science,* 6, 384–360, 1994.

Biddle, S., Sallis, J.F., and Cavill, N.A., *Young and Active? Young People and Health Enhancing Physical Activity. Evidence and Implication,* Health Education Authority, London, 1998.

Chu, N., Rimm, E., Wang, D., Liou, H., and Shieh, S., Clustering of cardiovascular disease risk factors among obese school children: The Taipei Children Heart Study, *Am. J. Clin. Nutr.*, 67, 1141–1146, 1998.

Clarke, R.W. and Lauer, R.M., Does childhood obesity track into adulthood? *Crit. Rev. Food Sci. Nutr.,* 33, 423–430, 1993.

Cooper, A.R. Page, A.S. Foster, L.J., and Qahwaji, D., Commuting to school: are children who walk more physically active? *Am. J. Prev. Med.,* 25, 273–276, 2003.

Dale, D., Corbin, C., and Dale, K., Restricting opportunities to be active during school time: Do children compensate by increasing physical activity levels after school? *Res. Q. Exerc. Sport,* 71, 240–248, 2000.

Department of the Environment, Transport and the Regions, *School Travel Strategies and Plans; A Best Practice Guide for Local Authorities,* The Stationery Office, London, 1999.

Department of the Environment, Transport and the Regions, *National Travel Survey: Update 1997/99,* The Stationery Office, London, 2000.

Ekelund, U., Aman, J., Yngve, A., Renman, C., Westerterp, K., and Sjostrom, M., Physical activity but not energy expenditure is reduced in obese adolescents: a case-control study, *Am. J. Clin. Nutr.,* 76, 935–941, 2002.

Hughes, J., Li, L., Chinn, S., and Rona, R., Trends in growth in England and Scotland, 1972 to 1994, *Arch. Dis. Child.,* 76, 182–189, 1997.

Jebb, S.A., Rennie, K.L., and Cole, T.J., Prevalence and demographic determinants of overweight and obesity among young people in Great Britain, *Int. J. Obes.,* 27 (Suppl. 1), S9, 2003.

Mallam, K.M., Metcalf, B.S., Kirkby, J., Voss, L.D., and Wilkin, T.J., Contribution of timetabled physical education to total physical activity in primary school children: cross sectional study, *BMJ,* 327, 592–593, 2003.

Morrow, J.R. Jr. and Freedson, P.S., Relationship between habitual physical activity and aerobic fitness in adolescents, *Pediatr. Exerc. Sci.,* 6, 315–319, 1994.

Must, A. and Strauss, R.S., Risks and consequences of childhood and adolescent obesity, *Int. J. Obes. Relat. Metab. Disord.,* 23 (Suppl. 2), S31–S37, 1999.

Ogden, C. Flegal, K., Carroll, M., and Johnson, C., Prevalence and trends in over-weight among US children and adolescents, 1999–2000 *JAMA,* 288, 1728–1732, 2002.

Pate, R.R., Freedson, P.S., Sallis, J.F., Taylor, W.C., Sirard, J., Trost, S.J., and Dowda, M., Compliance with physical activity guidelines: Prevalence in a population of children and youth, *Ann. Epidemiol.,* 12, 303–308, 2002.

Prentice, A.M. and Jebb, S.A., Obesity in Britain: gluttony or sloth? *BMJ,* 311, 437–439, 1995.

Riddoch, C.J., Anderson, L.B., Wedderkopp, N., Harro, M., Klasson-Heggebø, L., Sardinha, L.B., Cooper, A.R., and Ekelund, U., Physical activity levels and patterns of 9 and 15 year old European children, *Med. Sci. Sports Exerc.,* 36, 86–92, 2004.

Rowland, D., DiGuiseppi, C., Gross, M.E.A., and Roberts, I., Randomised controlled trial of site specific advice on school travel patterns, *Arch. Dis. Child.,* 88, 8–11, 2003.

Schwimmer, J.B., Burwinkle, T.M., and Varni, J.W., Health-related quality of life of severely obese children and adolescents, *JAMA,* 289, 1813–1819, 2003.

Sleap, M. and Warburton, P., Physical activity levels of 5–11-year old children in England: cumulative evidence from three direct observation studies, *Int. J. Sports Med.,* 17, 248–253, 1996.

Sorof, J., Poffenbarger, T., Franko, K., Bernard, L., and Portman, R., Isolated systolic hypertension, obesity, and hyperkinetic hemodynamic states in children. *J. Pediatr.,* 140, 660–666, 2002.

Sproston, K. and Primatesta, P. (Eds.), *Health Survey for England 2002: The Health of Children and Young People,* The Stationery Office, London, 2003.

Torok, K., Szelenyi, Z., Porszasz, J., and Molnar, D., Low physical performance in obese adolescent boys with metabolic syndrome, *Int. J. Obes. Relat. Metab. Disord.*, 25, 966–970, 2001.

Trost, S.G., Kerr, L.M., Ward, D.S., and Pate, R.R., Physical activity and determinants of physical activity in obese and non-obese children, *Int. J. Obes. Relat. Metab. Disord.*, 25, 822–829, 2001.

Trost, S.G., Pate, R.R., Freedson, P.S., Sallis, J.F., and Taylor, W.C., Using objective physical activity measures with youth: how many days of monitoring are needed? *Med. Ser. Sports Exerc.*, 32, 426–31, 2000.

Trost, S.G., Sirard, J.R., Dowda, M., Pfeiffer, K.A., and Pate, R.R., Physical activity in overweight and non-overweight preschool children, *Int. J. Obes. Relat. Metab. Disord.*, 27, 834–839, 2003.

Twisk, J.W.R., Physical activity guidelines for children and adolescents. A critical review, *Sports Med.*, 31, 617–627, 2001.

Twisk, J.W.R., Mellenbergh, G.J., van Mechelen, W., et al., Tracking of biological and lifestyle cardiovascular risk factors over a 14 year period, *Am. J. Epidemiol.*, 145, 888–898, 1997.

World Health Organization, *World Health Report 2002: Reducing Risk-Promoting Healthy Life*, available at www.who.int/whr/en, 2002.

10

INTERACTIONS AMONG PHYSICAL ACTIVITY, FOOD CHOICE, AND APPETITE CONTROL: HEALTH MESSAGES IN PHYSICAL ACTIVITY AND DIET

J. E. Blundell, N. A. King, and E. Bryant

CONTENTS

PHYSICAL ACTIVITY AND APPETITE: POSSIBLE SCENARIOS

It is often believed that physical activity is a poor strategy for losing weight since the energy expended will drive up hunger and food intake to compensate for the energy deficit incurred. However, evidence from studies in both normal-weight and obese adults shows that interventions of periods of exercise generate little or no *immediate* effect on levels of hunger or daily energy intake (EI) (see Blundell and King, 1999; and Blundell et al., 2003, for a review). However, less optimistically, there is no downregulation of appetite in response to periods of enforced sedentariness (Murgatroyd et al., 1999). Therefore, contrary to much opinion, evidence indicates a rather loose coupling between physical activity and appetite (food intake). Although a majority of this evidence comes from studies with adults, Moore et al. (2002) showed no immediate compensation for imposed exercise in 9- to 10-year-old girls. Moreover, in young people, (4 to 18 years), sedentary lifestyles are associated with a poor control over the intake of savory snacks (Rennie and Jebb, 2003). These data are consistent with results indicating that active individuals show better compensation for dietary energy loading than sedentary people (Long et al., 2002). Therefore, physical activity seems to enhance the sensitivity of satiety signals and improves appetite control through better meal to meal regulation.

The weak coupling between activity and food intake generates an optimistic view of the role of exercise in weight loss and weight control because it suggests that appetite is not automatically driven up to compensate for activity-induced energy expenditure. However, immediate or short-term effects may not be a good predictor of what happens over longer periods. Moreover, physical activity often produces disappointing effects on body weight. These could arise from inappropriate food choices and a feeling that food self-reward is justified or from misjudgments about the energy cost of physical activity (calories expended) relative to the rate of eating-induced intake (calories consumed). When physical activity is associated with high-fat, energy-dense foods, the beneficial effects of activity on energy balance can be reversed (King and Blundell, 1995; Tremblay et al., 1994). An increase in physical activity does not automatically protect against inappropriate food choice. Neither does long-term physical activity automatically improve macronutrient food choices (Donnelly et al., 2003). Therefore, food choice must be controlled independently of taking up exercise.

Although acute studies show that exercise does not immediately drive up food intake, it has recently been demonstrated that compensation does begin to occur if daily physical activity persists. Over periods of up to 16 days, partial compensation occurs so that increased eating accounts for approximately 30% of the increase in energy expended (Stubbs et al., 2002a). However, the existence of considerable individual differences indicates that for some individuals the compensation is approximately 100%.

WHAT ARE THE MAJOR HEALTH MESSAGES?

Importantly, physical activity has the potential to be a successful method of obesity prevention, but only if there is compliance with the prescribed amount, together with judicious control over food choice that involves selection of low- to medium-energy dense diet. However, the message must be accompanied by a warning that the delivery of the message does not ensure its implementation. Initially, there must be a realistic appreciation of the energy cost of physical activity compared with the energy contained in food. The widespread overestimation of the amount of energy used up by exercise, coupled with the underestimation of the amount of energy consumed in foods, generates a misleading impression of the amount of behavioral control required for energy balance and weight control. Second, it should be recognized that some individuals have the capacity to benefit from exercise more than others. The early identification of individuals resistant or susceptible to the beneficial effects of exercise on body weight should be an important step in optimizing the effects of exercise. Those individuals more resistant to exercise will need additional or alternative strategies to help them reach a target of a more healthy weight. We argue that there needs to be a much greater understanding of individual human variability — in children and adults — before the appropriate health messages can be effective.

THE EFFECT OF INCREASED PHYSICAL ACTIVITY ON ENERGY INTAKE

There are many examples in the literature of food-induced reductions in energy intake (EI), for example, skipping a meal, which leads to compensatory increases in hunger and food intake, causing an acute energy deficit (Delargy et al., 1995; Hubert et al., 1998; Green et al., 1994). Therefore, it is logical to infer that, by creating an energy deficit with energy expenditure (EE), physical activity will have a similar effect on energy balance. Most of the evidence to examine the relationship between physical activity and food intake comes from short-term intervention studies in adults, which assess the effects of deliberately imposed acute bouts of exercise on subsequent food (energy) intake. Contrary to the widespread belief, there is no immediate compensatory increase in hunger and food intake (Imbeault et al., 1997; King et al., 1994; King et al., 1995; King et al., 1996; King et al., 1997; Kissileff et al., 1990; Lluch et al., 1998; Reger et al., 1984; Thompson et al., 1988; Westerterp-Plantenga et al., 1997). Recently, this phenomenon was also demonstrated in 9- to 10-year-old girls (Moore et al., 2002). Therefore, overall, the body of evidence points to a loose coupling between exercise-induced EE and EI (for reviews, see Blundell et al., 2003; Blundell and King, 1998; King et al., 1997; King, 1998).

One criticism is that these short-term studies fail to track EI for a sufficiently long period following the increased physical activity interventions and that the exercise-induced increment in EE is not large enough to stimulate appetite. However, even with a high dose of exercise (gross exercise-induced increase in EE = 4.6 MJ) in a single day and tracking EI for the following 2 days, there is no automatic compensatory rise in hunger and EI (King et al., 1997). Therefore, the evidence suggesting that acute exercise-induced negative energy balances (EB) is not compensated for by an increase in EI is fairly robust. One reason for this loose coupling is that the behavioral act of eating is held in place by environmental contingencies and short-term postingestive physiological responses arising from eating itself. Exercise -induced changes in postabsorptive physiology (energy metabolism) appear to have only a weak influence on eating behavior.

Although overall the evidence from the acute studies indicates that EI is not immediately driven up by increased physical activity, we have recently demonstrated that partial compensation starts to occur if physical activity persists. The evidence confirmed that, over a period of 7 days (Stubbs et al., 2002a; Stubbs et al., 2002b) and 14 days (Stubbs et al., 2004), EI did not remain constant following the marked elevation of EE. Findings from the 14-day exercise intervention suggested that, on average, subjects compensated for ~30% of the exercise-induced energy deficit (Stubbs et al., 2004). However, there was considerable variation in the extent of compensation between individuals such that some compensated completely (100%) for the increase in EE. It is likely that some individuals have the capacity to benefit from exercise more than others.

WHY IS EXERCISE NOT ALWAYS SUCCESSFUL AT INDUCING WEIGHT LOSS?

The robust evidence pointing to a loose coupling between exercise-induced EE and EI generates an optimistic view of the role of exercise in weight control because it dispels the belief that appetite will be automatically driven up to compensate for activity-induced increases in EE. Therefore, from a practical perspective, exercise should be a good technique to bring about weight loss. However, physical activity does not always result in successful weight control and often produces disappointing effects on body weight. There may be a number of reasons why this is the case. For example, failure to maintain a 100% compliance with the exercise regime and reduction in physical activity during nonexercise time (recovery periods) could both contribute to a lack of weight loss. Inappropriate

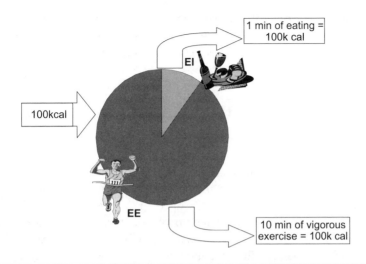

Figure 10.1 A Schematic Diagram Outlining the Biological Mismatch Between the Rate (Low) of Energy Expenditure (EE; Exercise) Compared With the Rate (High) of Energy Intake (EI; Eating).

food choices and allowance of food rewards, as well as misjudgments about the rate of eating-induced intake (calories consumed) relative to the energy cost of physical activity (calories expended), could also jeopardize the outcome. Such behavioral adaptations would be superimposed upon any metabolic adjustment (e.g., reduction in resting metabolic rate), thereby resulting in an energy savings.

Some individuals make poor evaluations of the amount of energy that can be expended during exercise and the amount that can be ingested during eating. For a fixed level of energy, the duration of exercise (expenditure) has to be significantly greater compared with the duration of eating (intake) (see Figure 10.1). For example, to expend 600 kcal, an individual of moderate fitness [i.e., maximal oxygen uptake (VO$_2$max) of 3 l/min] would have to exercise for approximately 60 minutes at 75% VO$_2$max. However, any individual (independent of aerobic fitness) could ingest 600 kcal of food energy in the form of an energy-dense snack (e.g., a Danish pastry or a couple of doughnuts) in 3 to 4 minutes. Consequently, individuals must be informed about the possible "mismatch" between the *rate* of energy expenditure (low) and *rate* of energy intake (high).

Several studies have demonstrated that the beneficial effects of exercise on energy balance can be completely reversed when physical activity is combined with high-fat, energy-dense foods and diets (King et al., 1994; Tremblay et al., 1994; Murgatroyd et al., 1999). An increase in physical activity does not automatically protect against inappropriate food choice. Therefore, physical activity should not be viewed as an opportunity to abandon any restraint over eating or to indulge excessively on available foods.

HOW DOES SEDENTARINESS AFFECT ENERGY INTAKE?

Although the loose coupling between exercise-induced EE and EI has positive implications for weight control for increases in EE, unfortunately it has negative implications for decreases in EE. When EE decreases in individuals who become sedentary, energy intake is not drawn downward to a new lower level in equilibrium with the reduced EE. Recent findings have confirmed that there is no downregulation of appetite and that EI remains at some stable preferred level when EE is reduced. Inactivity-induced reductions in EE occur in the natural, free-living environment due to a variety of reasons (e.g., injury, increase in energy-saving devices, increased car use). Less natural, but more accurate, imposed reductions in EE can be achieved when individuals reside in a whole body calorimeter. A short-term study (Murgatroyd et al., 1999) and a medium-term study (Stubbs et al., 2004) have demonstrated that a reduction in EI is not induced to compensate for activity-induced reductions in EE; hence a positive energy balance ensues. Therefore, physical inactivity does not reduce food intake, and, bearing in mind that eating tends to be a sedentary activity, inactivity could even increase food intake. Indeed, a recent study showed that time spent inactive was associated with a higher savory snack intake (Rennie and Jebb, 2003). Thus sedentariness could be a risk factor for two reasons: a natural decrease in EE and an increase in EI due to consumption of energy dense foods (see Figure 10.2).

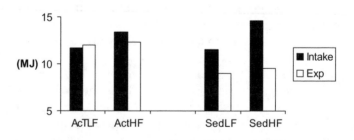

Figure 10.2 Mean Energy Intake and Energy Expenditure (Exp) (in MJ) During Days of Imposed Activity (Act) and Sedentariness (Sed) in Subjects That at Low-Fat (LF) or High-Fat (HF) Diets. Energy Intake is not Downregulated in Response to the Imposed Sedentariness Despite a Reduction in Energy Expenditure. (Based on data from Murgatroyd et al., *Int. J. Obes. Relat. Metab. Disord.*, 23, 1269–1275, 1999).

KEY CONCEPT: INDIVIDUAL VARIABILITY IN THE RESPONSE TO EXERCISE

It has been noted earlier that there is a considerable variability in the way in which individuals respond to an imposed exercise regime. Some sustain an energy deficit and lose weight. However, in others, the energy deficit serves as a drive to increase food consumption or to decrease physical activity during the nonexercise periods. It is also highly probable that some individuals fail to benefit from exercise because they fail to comply with the instructions. One reason for this could be that individuals do not derive a significant level of pleasure or satisfaction from engagement in exercise per se. In other words, a particular type of conservative physiology creates a resistance to physical activity and consequently, an energy deficit, by lowering the hedonic value of exercise. This situation is analogous to, but can be contrasted with, the way in which some people find eating food a much more pleasurable activity than others. Individuals susceptible to weight gain report that food has a greater hedonic appeal than that reported by people resistant to weight gain, which thus leads to more food consumption (Blundell and Le Noury, 2003). Similarly, we have recently reported striking individual differences in the degree of positive affective state generated by periods of exercise (Purslow et al., 2003). Indeed, some individuals report that exercise actually lowers their positive mood. Accordingly, it is appropriate to examine further the extent to which individual differences in the subjective feelings generated by exercise contribute to the overall effectiveness of exercise as a controller of body weight. This factor will be equally relevant to adults and children.

INDIVIDUAL DIFFERENCES IN THE REWARD SENSITIVITY TO FOOD AND EXERCISE

In considering this issue, it is appropriate to think of the "hedonic" dimension of behavioral action; that is, the degree of pleasure induced and the power of the action to induce reinforcement or reward. Individuals are more likely to vary in their extent to which an action activates reward processes or pathways. Indeed, sensitivity to reward may reflect an innate characterological trait (Davis and Woodside, 2002), where some individuals are more tuned for reward from activities such as exercise and eating than others. Evidence indicates that vigorous exercise and eating behaviors, particularly food restriction, stimulate brain substrates that are associated with reward and dependence (Bergh and Soderstern, 1996; Lett et al., 1996). A stimulation of the hypothalamic-pituitary-adrenal (HPA) axis is elicited by these behaviors; beta-endorphins are released, which activate the dopaminergic neurons in the mesolimbic brain structures, which, in turn, activate the common reward pathway.

The capacity for reward is proposed to be a normally distributed trait within the population (Meehl, 1975). Indeed, research has highlighted that genetically determined differences exist in dopamine availability in areas of the brain associated with reward and that a lower availability is linked to a higher risk of addiction (Sabol et al., 1999). Blum et al. (1997) suggested the idea of a "reward deficiency syndrome," whereby deprivation of the brain's reward chemicals, leading to an individual's inability to derive pleasure from everyday activities, can cause a person to seek out compensatory behaviors as a way of self-medicating negative affective states (Lett et al., 1996; Carton et al., 1994). On this premise, it is assumed that food and exercise can be used, in a similar fashion to addictive drugs, as a way to stimulate brain dopamine levels.

Recently, Volkow et al. (1999) suggested a simple optimal level, inverted U relationship between hedonic inclination and dopamine activation, in which either too little or too much is aversive. Therefore, if an individual has a low sensitivity to reward (low dopamine availability) and they exercise or eat palatable foods, they will boost their low-activity dopamine system, thus exerting positive subjective effects without increasing the activation beyond optimal levels. This allows a higher amount of the compensatory behavior to be endured before the optimal activation levels are reached and the behavior becomes aversive. Individuals with high sensitivity to reward (high dopamine levels), however, experience positive subjective effects from just a small stimulation of dopamine, and thus a large increase in dopamine levels, arising from, for example, too much palatable food or exercise, will be experienced as aversive (Wang et al., 2002).

Obesity could be due to individuals having a low sensitivity to reward; therefore, individuals could self-medicate themselves by overeating palatable foods. However, because overeating can alter brain mechanisms in a way similar to addictive drug abuse (Gamberino and Gold, 1999), hedonically driven overeating may cause the mesolimbic system to compensate for overstimulation by downregulating dopamine availability (Davis et al., 2004). Thus, the behavior itself may induce an anhedonic state (Zacharko, 1994) in individuals who were nearer the hedonic end of the continuum, leading to an increased salience of the behavior, even if the reward of consumption diminishes (Robinson and Berridge, 2003).

Similarly, Davis and Woodside (2002) found that strenuous exercise can have the same effect as overconsumption; because this can stimulate dopamine and the endogenous opioid system, it may provide a mechanism for mood regulation; however, this could eventually generate an anhedonic state (Zacharko, 1994). Subsequently, more exercise would be needed to gain the same level of hedonic tone.

One consequence of this analysis is that a single type of exercise prescription is unlikely to work for everyone, since individuals vary in the degree to which they can experience a positive subjective effect. An important health message is that, for those individuals who fail to get a reward "buzz" from exercise, additional strategies should be imposed in order to deliver a positive state.

DOES PHYSICAL ACTIVITY AFFECT APPETITE SENSITIVITY?

Appetite sensitivity is how precise the appetite system is in detecting when the body has consumed enough energy. This can be achieved through enhancement of satiation signals or by compensating for a previous energy load in the next meal. Evidence suggests that exercisers, or habitually physically active individuals, are better able to regulate their food intake and energy balance because of an increased appetite sensitivity. Mayer et al. (1956) were the first to highlight the flaw in the belief that energy intake functions in such a way that it automatically increases after an increase in energy expenditure and decreases after a reduction in energy expenditure. In a study of workers in West Bengal, Jean Mayer found that energy intake only increases with activity within a certain zone of activity ("normal activity"); below this range ("sedentary zone"), a decrease in activity is not associated with a decrease in food intake. Rather it is associated with an increase in food intake together with an increase in body weight.

More recently, Long et al. (2002) demonstrated that habitual exercisers have an increased accuracy of short-term regulation of energy intake in comparison to nonexercisers. In this study, participants were given either a low- or high-energy preload for lunch and were then asked to eat ad libitum from a test meal buffet (see Figure 10.3). Energy intake from the buffet did not significantly differ following the two preloads in the nonexercise group,

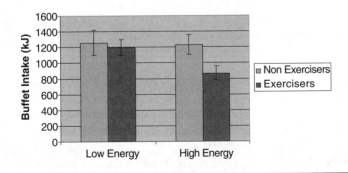

Figure 10.3   Mean Energy Intake (in kJ) in Habitual Exercisers and Nonexercisers Following Low-Energy and High-Energy Preloads. Habitual Exercisers Compensated for the High-Energy Load, Whereas the Non-Exercisers did not. (Based on data from Long et al., *Br. J. Nutr.*, 87, 517–523, 2002).

showing a compensation of only 7%. On the other hand, the habitual exercisers reduced their energy intake after the high-energy preload compared with the low-energy preload, exhibiting a 90% compensation. No differences between groups were observed for subjective ratings of hunger or satiety; thus the differences, which became apparent between exercisers and nonexercisers, were due to differences in intake. This suggests that exercise may increase sensitivity to satiety signals as opposed to hunger signals and that energy intake could be more successfully controlled through the size of an eating period. These data complement previous work in which it was proposed that obese individuals are less sensitive than lean individuals to short-term, postingestive satiety signals (Spiegel et al., 1989). Consequently, both sedentary individuals and overweight people are characterized by low sensitivity in the appetite control system. In other words, the system does not respond accurately to interventions that provoke changes in energy balance.

Although the relationship between habitual exercise and appetite is becoming clearer, it is also recognized as being more complicated, and the mechanisms are not yet well defined. Evidence suggests that a possible mechanism through which habitual exercise could promote regulation of appetite is through insulin sensitivity. Exercise is known to increase insulin sensitivity (e.g., Andersson et al., 1991; Donnelly et al., 2000; Poehlman et al., 2000), and insulin sensitivity is known to be involved in satiety induced by particular foods (Holt et al., 1992) and in the compensatory response to high-energy loads (Speechly and Buffenstein, 2000). A further mechanism by which exercise could affect appetite is through altering gut peptide action. For example, cholecystokinin (CCK) is implicated in the short-term regulation of appetite, and levels of CCK have been shown to rise after exercise (Bailey et al., 2001). However, the effects of exercise on other gut peptides, such as PYY, GLP-1, and ghrelin, involved in the episodic regulation of appetite are to date unknown. In addition, habitual exercising could influence the level of "tonic" appetite signals such as leptin or adiponectin, which reflect the size of body fat stores (Halford and Blundell, 2000).

SUMMARY

It is clear that there is not a simple relationship between hunger (and the pattern of eating) and perturbations in EE induced by adjustments in physical activity. Whether EE is increased (imposed exercise) or decreased, eating habits continue in their habitual form. There is strong evidence for a loose coupling between activity-induced EE and EI. This has optimistic implications for the use of exercise in weight control but is problematic for a nation that is becoming increasingly sedentary. However, there is considerable interindividual variability in the response to exercise that has not yet been explored. These individual differences include a variable hedonic responsivity to exercise (making it more, or less, pleasurable) and a variation in the sensitivity of the appetite control system. Taken together, these two factors — hedonic responsivity and appetite sensitivity — will help to determine whether any individual — child or adult — is likely to benefit from undertaking exercise. This individual variability contains a clear health message. It is abundantly clear that the capacity to benefit from exercise involves much more than a simple coercive instruction to do more. This message is unlikely to be effective unless it is combined with a more empathic approach to biological and psychological differences between human beings.

REFERENCES

Andersson, B., Xuefan, X., Rebuffe-Scrive, M., Terning, K., Krotkiewski, M., and Bjorntorp, P., The effects of exercise training on body composition and metabolism in men and women, *Int. J. Obes. Relat. Metab. Disord.*, 15, 75–81, 1991.

Bailey, D.M., Davies, B., Castell, L.M., Newsholme, E.A., and Calam, J., Physical exercise and normobaric hypoxia: independent modulators of peripheral cholecystokinin metabolism in man, *J. Appl. Physiol.*, 90, 105–113, 2001.

Bergh, C. and Sodersten, P., Anorexia nervosa, self-starvation and the reward of stress, *Nat. Med.*, 2, 21–22, 1996.

Blum, K. and Noble, E.P., et al., Eds., *Handbook of Psychiatric Genetics,* CRC Press, Boca Raton, 1997, pp. 311–327.

Blundell, J.E. and King, N.A., Effects of exercise on appetite control: loose coupling between energy expenditure and energy intake, *Int. J. Obes. Relat. Metab. Disord.*, 22, 1–8, 1998.

Blundell, J.E. and Le Noury, J., Appetite control and palatability of food in humans: does the pleasure of eating lead to obesity, in *Progress in Obesity Research 9,* Medeiros-Neto, G., Halpern, A., and Bouchard, C., Eds., pp. 822–825, 2003.

Blundell, J.E., Stubbs, R.J., Hughes, D.A., Whybrow, S., and King, N.A., Cross talk between physical activity and appetite control: does physical activity stimulate appetite? *Proc. Nutr. Soc.*, 62, 651–661, 2003.

Carton, S., Jouvent, R., Widlocher, D., et al., Nicotine dependence and motives for smoking in depression, *J. Substance Abuse.* 6, 67–76, 1994.

Davis, C. and Woodside, D.B., Sensitivity to the rewarding effects of food and exercise in the eating disorders, *Compr. Psychiatry,* 43, 189–194, 2002.

Davis C., Strachan S., and Berkson M., Sensitivity to reward: implications for overeating and overweight, *Appetite.* 42, 125–140, 2004.

Delargy, H.D., Burley, V.J., Sullivan, K.R., Fletcher, R.J., and Blundell., J.E., Effects of different soluble:insoluble fibre ratios at breakfast on 24–h pattern of dietary intake and satiety, *Eur. J. Clin. Nutr.*, 49, 754–766, 1995.

Donnelly, J.E., Jacobsen, D.J., Heelan, K., Seip, R., and Smith, S., The effects of 18 months of intermittent vs. continuous exercise on aerobic capacity, body composition, and metabolic fitness in previously sedentary, moderately obese females, *Int. J. Obes. Relat. Metab. Disord.*, 24, 566–572, 2003.

Gamberino, W.C. and Gold, M.S., Neurobiology of tobacco smoking and other addictive behaviours, *Psychiatr. Clin. North Am.*, 22, 301–312, 1999.

Green, S.M., Burley, S.M., and Blundell, J.E., Effect of fat- and sucrose-containing foods on the size of eating episodes and energy intake in lean males: potential for causing overconsumption, *Eur. J. Clin. Nutr.*, 48, 547–555, 1994.

Halford, J.C.G. and Blundell, J.E. (2000) Differing roles for serotonin and leptin in appetite regulation, *Ann. Med.*, 32, 222–232, 2000.

Holt, S., Brand, J., Soveny, C., and Hansky, J., Relationship of satiety to postprandial glycaemic, insulin and cholecystokinin responses, *Appetite*, 18, 129–141, 1992.

Hubert, P., King, N.A., and Blundell, J.E., Uncoupling the effects of energy expenditure and energy intake: appetite response to short-term energy deficit induced by meal omission and physical activity, *Appetite*, 31, 9–19, 1998.

Imbeault, P., Saint-Pierre, S., Almeras, N., and Tremblay, A., Acute effects of exercise on energy intake and feeding behaviour, *Br. J. Nutr.*, 77, 511–521, 1997.

King, N.A., Burley, V.J., and Blundell, J.E., Exercise-induced suppression of appetite: effects on food intake and implications for energy balance. *Eur. J. Clin. Nutr.*, 48, 715–724, 1994.

King, N.A., Blundell, J.E. (1995). High-fat foods overcome the energy expenditure due to exercise after cycling and running, *Eur. J. Clin. Nutr.*, 49, 114–123, 1995.

King, N.A., Snell, L., Smith, R.D., and Blundell, J.E. Effects of short-term exercise on appetite response in unrestrained females, *Eur. J. Clin. Nutr.*, 50, 663–667, 1996.

King, N.A., Tremblay, A., and Blundell, J.E., Effects of exercise on appetite control: implications for energy balance, *Med. Sci. Sports Exerc.*, 29, 1076–1089, 1997.

King, N.A., Lluch, A., Stubbs, R.J., and Blundell, J.E., High dose exercise does not increase hunger or energy intake in free living males, *Eur. J. Clin. Nutr.*, 51, 478–483, 1997.

King, N.A., The relationship between physical activity and food intake, *Proc. Nutr. Soc.*, 57, 1–9, 1998.

Kissileff, H.R., Pi-sunyer, X.F., Segal, K., Meltzer, S., and Foelsch, P.A., Acute effects of exercise on food intake in obese and non-obese women, *Am. J. Clin. Nutr.*, 52, 240–245, 1990.

Lett, B.T., Grant, V.L., and Gaborko, L.L., A small amount of wheel running facilitates eating in nondeprived rats, *Behav. Neurosci.*, 110, 1492–1495, 1996.

Lluch, A., King, N.A., and Blundell, J.E., Exercise in dietary restrained women: no effect on energy intake but change in hedonic ratings, *Eur. J. Clin. Nutr.*, 52, 300–307, 1998.

Long, S.J., Hart, K., and Morgan, L.M., The ability of habitual exercisers to influence appetite and food intake in response to high- and low-energy preloads in man, *Br. J. Nutr.*, 87, 517–523, 2002.

Mayer, J., Roy, P., and Mitra, K.P., Relation between caloric intake, body weight, and physical work: studies in an industrial male population in West Bengal, *Am. J. Clin. Nutr.*, 4, 169–174, 1956.

Meehl, P., Hedonic capacity: some conjectures, *Bull. Menninger Clin.*, 39, 295–307, 1975.

Moore, M.S., Dodd, C.D., Welsman, J.R., and Armstrong, N., Lack of short-term compensation for imposed exercise in 9–10 year-old girls, *Proc. Nutr. Soc.*, 61, 156A, 2002.

Murgatroyd, P.R., Goldberg, G.R., Leahy, F.E., Gilsenan, M.B., and Prentice, A.M., Effects of inactivity and diet composition on human energy balance, *Int. J. Obes. Relat. Metab. Disord.*, 23, 1269–1275, 1999.

Poehlman, E.T., Dvorak, R.V., DeNino, W.F., Brochu, M., and Ades, P.A., Effects of resistance training and endurance training on insulin sensitivity in nonobese, young women: a controlled randomized trial, *J. Clin. Endocrinol. Metab.*, 85, 2463–2468, 2000.

Purslow, L., King, N.A., and Blundell, J.E., Heart rate and mood change during exercise: key to compliance, *Int. J. Obes. Relat. Metab. Disord.*, 27, S140, 2003.

Reger, W.E., Allison, T.A., and Kurucz, R.L., Exercise, post-exercise metabolic rate and appetite, *Sport Health Nutr.*, 2, 115–123, 1984.

Rennie, K.L. and Jebb, S.A., Sedentary lifestyles are associated with being overweight and consumption of savoury snacks in young people (4–18) years, *Proc. Nutr. Soc.*, 62, 83A, 2003.

Robinson, T.E. and Berridge, K.C., Addiction, *Annu. Rev. Psychol.*, 54, 10.1–10.29, 2003.

Sabol, S.Z., Nelson, M.L., Fisher, C., et al., A genetic association for cigarette smoking behaviour, *Health Psychol.*, 18, 7–13, 1999.

Speechly, D.P. and Buffenstein, R., Appetite dysfunction in obese males: evidence for role of hyperinsulinaemia in passive overconsumption with a high fat diet, *Eur. J. Clin. Nutr.*, 54, 225–233, 2000.

Spiegel, T.A., Shrager, E.E., and Stella, E., Responses of lean and obese subjects to preloads, deprivation and palatability, *Appetite*, 13, 45–69, 1989.

Stubbs, R.J., Sepp, A., Hughes, D.A., Johnstone, A.M., King, N., Horgan, G.W., and Blundell, J.E., The effect of graded levels of exercise on energy intake and balance in free-living women, *Int. J. Obes. Relat. Metab. Disord.*, 26, 866–869, 2002a.

Stubbs, R.J., Sepp, A., Hughes, D.A., Johnstone, A.M., Horgan, G.W., King, N., and Blundell, J., The effect of graded levels of exercise on energy intake and balance in free-living men, consuming their normal diet, *Eur. J. Clin. Nutr.*, 56, 129–140, 2002b.

Stubbs, R.J., Hughes, D.A., Johnstone, A.M., Whybrow, S., Horgan, G.W., King, N., and Blundell, J., Rate and extent of compensatory changes in energy intake and expenditure in response to altered exercise and diet composition in humans, *Am. J. Physiol. Regul. Integr. Comp. Physiol.*, 286, R350–R358, 2004.

Stubbs, R.J., Hughes, D.A., Ritz, P., Johnstone, A.M., Horgan, G.W., King, N., and Blundell, J.E., A decrease in physical activity affects appetite, energy and nutrient balance in lean men feeding ad libitum, *Am. J. Clin. Nutr.*, 79, 62–69, 2004.

Thompson, D.A., Wolfe, L.A., and Eikelboom, R., Acute effects of exercise intensity on appetite in young men, *Med. Sci. Sports Exerc.*, 20, 222–227, 1988.

Tremblay, A., Almeras, N., Boer, J., Kranenbarg, E.K., and Despres, J.P., Diet composition and postexercise energy balance, *Am. J. Clin. Nutr.*, 59, 975–979, 1994.

Volkow, N.D., Wang, G.J., Fowler, J.S., et al., Prediction of reinforcing responses to psychostimulants in humans by brain dopamine D2 receptor levels, *Am. J. Psychiatry*, 156, 1440–1443, 1999.

Wang, G.J., Volkow, N.D., and Fowler, J.S., Dopamine's role in motivation for food in humans: implications for obesity, *Expert Opinion*, 6, 601–609, 2002.

Westerterp-Plantenga, M.S., Verwegen, C.R.T., Ijedema, M.J.W., Wijckmans, N.E.G., and Saris, W.H.M., Acute effects of exercise or sauna on appetite in obese and non-obese men, *Physiol. Behav.*, 62, 1345–1354, 1997.

Zacharko, R.M., Stressors, the mesolimbic system, and anhedonia: implications for PTSD, in *Catecholamine Function in Posttraumatic Stress Disorder: Emerging Concepts. Progress in Psychiatry*, Murburg, M.M., Ed., American Psychiatric Press, Washington, DC, no. 42, pp. 99–130, 1994.

11

PERINATAL INFLUENCES ON CHILDHOOD OBESITY RISK

David B. Dunger and Ken K. Ong

CONTENTS

INTRODUCTION

Size at birth and early postnatal weight gain are closely related to neonatal and childhood survival (Karn and Penrose, 1951), and in recent years there have been many studies suggesting that these may also predict much longer term risks of diseases in adult life, such as Type 2 diabetes and cardiovascular disease (Barker et al., 1989, Hales et al., 1991). The most commonly reported pattern of association has been in subjects with relatively lower birth weights and who subsequently become overweight or obese either during childhood or adult life (Eriksson et al., 1999). Characteristically, such subjects tend to have reduced muscle mass and increased central adiposity in association with insulin resistance (Hediger et al., 1998). These findings have led to the concept of a "thrifty phenotype," in which early exposure to poor nutrition leads to permanent changes in insulin metabolism and body fat distribution (Hales and Barker, 2001).

Risks in adults for diabetes and cardiovascular disease are particularly high among populations where, until relatively recently, nutrition was poor but now rates of obesity are increasing (Zimmet, 1999). Such observations have led to the proposition that "thrifty genotypes" may have evolved, which, through possible actions on insulin activity and fat deposition, may have provided some survival advantage during long periods of fasting or famine but are now detrimental to later health when such events are less common and food is more plentiful (Neel, 1962).

Given the strong relationship between adequate perinatal growth and the relatively high mortality rates during early human development (Karn and Penrose, 1951), it is likely that both "thrifty phenotypes" and "thrifty genotypes" may have evolved to protect early survival against conditions of poor nutrition but may now determine the future disease risks associated with current increases in childhood obesity.

ASSOCIATION BETWEEN SIZE AT BIRTH AND ADULT DISEASE

Since those early reports in historical birth cohorts of association between smaller size at birth and cardiovascular disease and adult onset diabetes, similar findings have been confirmed in diverse populations from different countries (Stein et al., 1996; Forsen et al., 1997). The findings persist when differences in gestational age are taken into account, and they appear to be independent of selection bias or potential confounding due to social class or smoking (Leon et al., 1998). Importantly, the birth weight associations are not confined to differences between the smallest vs. the other infants; rather, these relate to a continuum of variable risk throughout the whole range of birth weights. For example, the original studies in men born in Hertfordshire, U.K. between 1911 and 1930 found that those with above average birth weight had 24% lower standardized mortality rates from coronary artery disease compared with those with average birth weights (Barker et al., 1989). Indeed, in some populations, the birth weight-adult disease association appears to be "U-shaped," with the heaviest-born babies also having an increased long-term risk for disease (McCance et al., 1994; Lindsay et al., 2000). Recent longitudinal growth data, including early childhood growth, in subjects from Finland who went on to develop Type 2 diabetes in adult life showed that both larger and smaller birth weight patterns are associated with increased disease risk (Eriksson et al., 2003). It is likely that the long-term disease associations with larger birth weight may reflect the influence of maternal diabetes in promoting both larger birth size and conferring offspring diabetes risk (Dabelea et al., 2000).

Epidemiological studies indicate that size at birth and early weight gain predict the long-term risks for obesity and abnormal fat distribution. In a study of 300,000 19-year-old men exposed to the Dutch Famine between 1944 and 1945, there was a nearly twofold increase in obesity risk in those subjects whose mothers were exposed to famine during the first trimester of pregnancy (Ravelli et al., 1976). Gale et al. (2001) showed that, among 70- to 75-year-old men studied by dual energy x-ray absorptiometry scanning, low birth weight was associated with reduced lean tissue mass and greater body fat relative to current weight. The predisposition to adult disease conferred by low birth weight may therefore be related to excess fat deposition, in particular central fat, and the development of insulin resistance. One study, which used the gold standard euglycemic-hyperinsulinemic clamp assessment of insulin sensitivity in 70-year-old men, showed that the association between low birth weight and insulin resistance was only seen in the highest body-mass index (BMI) tertile group (McKeigue et al., 1998). Thus, it appears to be the transition from relatively lower birth weight to larger postnatal body size that confers disease risk. With increasing abundance of nutrition and rising rates of obesity, such a transition may occur at younger ages (Reilly et al., 1999; Bundred et al., 2001), and its effects are currently being explored in contemporary birth cohort studies.

SIZE AT BIRTH AND EARLY POSTNATAL WEIGHT GAIN

Growth data from large contemporary birth cohorts, such as the Avon Longitudinal Study of Pregnancy and Childhood (ALSPAC) (Golding et al., 2001), confirm much earlier observations in smaller studies (Tanner, 1986) that around 25% of all newborn infants will show a significant degree of postnatal rapid or "catch-up" growth (Ong et al., 2000). Such babies tend to be longer at birth with a larger head circumference relative to their birth weight and therefore show reduced adiposity at birth compared with other babies of the same birth weight. A further 25% of all newborn infants have relatively increased adiposity at birth and will show postnatal slow or "catch-down" growth (Ong et al., 2000). The remaining infants who do not show postnatal catch-up or catch-down growth are those who grow along the same weight and length centile positions and appear to more closely follow their genetic growth trajectory from birth, as indicated by consistent strength of correlation with their midparental target heights (Dunger et al., 1998). Catch-up and catch-down postnatal growth are most marked in terms of changes in adiposity and are largely seen within the first 12 months of life, although they may take up to 2 years to complete (Ong et al., 2002).

The large extent of early postnatal catch-up and catch-down growth suggests that wide variations in gains of adiposity may also occur during late pregnancy. In the ALSPAC birth cohort, we have shown that early postnatal catch-up and catch-down growth are closely related to maternal factors during pregnancy, such as mother's pregnancy weight gain, smoking during pregnancy, and parity (birth order). The effect of parity is particularly striking in that first babies are more likely to be restrained in utero, are thinner at birth, and show early postnatal catch-up growth, whereas offspring of subsequent pregnancies are more likely to show in utero growth enhancement with postnatal catch-down growth (Ong et al., 2002). There is some evidence to suggest that these postnatal patterns of weight gain are driven by satiety, as indicated by early feeding studies in infants by Ounsted (Ounsted and Sleigh, 1975) and by the finding of significant relationships between the levels of satiety hormone leptin in cord blood at birth and subsequent patterns of weight gain (Ong et al., 1999a). By the time early postnatal catch-up and catch-down growth are completed, the infants are closer to their genetic target size, as predicted by their parents' heights (Dunger et al., 1998; Tanner, 1986). Thus, during the early months of life, when feeding patterns are strongly influenced by the infant and growth is regulated by nutrition, inherent patterns of increased or decreased appetite and satiety may return the infant toward its genetic growth trajectory.

However, these patterns of early postnatal growth also appear to have more long-lasting effects. In the ALSPAC cohort, subjects who showed early catch-up growth became the heaviest of all children at the age of 5 years (Ong et al., 2000). This excess weight persisted in ALSPAC catch-up children at their recent 8 year follow-up (Figure 11.1), and similar observations have been made in large cohort studies in the U.S. and in the Seychelles (Stettler et al., 2002b; Stettler et al., 2002a). In addition to BMI (weight for height), in our studies those ALSPAC children who showed early catch-up growth also had increased abdominal circumference at age 5 years. In other populations, the transition from low birth weight to a normal or increased BMI during childhood has been associated with alterations in body composition: low birth weight has been associated with increased central fat deposition in children and adults (Law et al., 1992; Yajnik, 2000). In the Third National Health and Nutrition Examinations Survey (NHANES 3) 1988–1994, children born small for gestational age showed reduced lean tissue mass without reduction in fat mass, and thus they had a higher percent body fat (Hediger et al., 1998). Studies from Australia have also

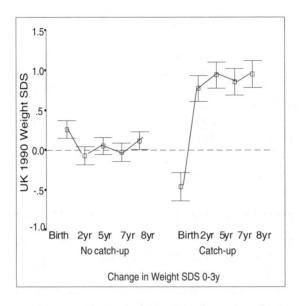

Figure 11.1   Rapid "Catch-up" Early Postnatal Weight Gain (Gain in Weight SD Score > +0.67 Between 0 and 3 Years) has a Persisting Effect on Increased Subsequent Childhood Weight from 5 to 8 Years. (From the ALSPAC Children in Focus Cohort).

reported this association between low birth weight, increased current weight, and increased central fat deposition (Garnett et al., 2001).

In a study from Pune, Indian children with relatively lower birth weights had increased fasting insulin levels at age 8 years, and similar findings have been reported in other populations (Bavdekar et al., 1999). Girls from Barcelona, Spain, who had relatively lower birth weights and who showed postnatal catch-up growth, became insulin resistant with increased central adiposity, although they may not necessarily be obese as assessed by standard BMI cut-offs (Ibanez et al., 2003a). In the ALSPAC cohort, we recently reported that postnatal catch-up growth is associated with insulin resistance at age 8 years, and that for this outcome the critical timing of catch-up is within the first 2 to 3 years of life (Ong et al., 2004b). In a recent case-control study from Chile, infants with low birth weight for gestational age who showed early postnatal catch-up growth already had higher fasting insulin levels by at the age of 1 year than infants of higher birth weights, even though the low birth weight infants had not yet attained the same weight at 1 year as the other infants (Soto et al., 2003).

These data suggest that low birth weight followed by early postnatal catch-up could be a risk factor for later obesity and disease risk and that the development of insulin resistance and increased central adiposity may be a very early feature of this growth pattern. These findings are comparable to results generated by animal models where prenatal growth restriction, followed by postnatal *ad libitum* feeding can result in insulin resistance and diabetes (Hales and Barker, 2001). However, in human childhood studies, links between insulin resistance and birth weight are often only evident in the largest BMI tertile during childhood. Thus, it is yet to be determined whether it is the interaction between size at birth and postnatal weight gain or the prenatal nutritional exposure, which is paramount in the risk for obesity and insulin resistance.

THRIFTY GENOTYPES AND OBESITY RISK

The thrifty genotype hypothesis was originally proposed by Neel in 1962 to explain the remarkably high prevalence of Type 2 diabetes in recently Westernized, previously undernourished populations (Neel, 1962). Although obesity rates have increased dramatically in developed countries and urban populations, the risk of obesity-related disease appears to be disproportionately distributed, with very high rates among subjects from ethnic populations who had experienced poor nutrition until relatively recently in their history (WHO-Expert-Consultation, 2004; Zimmet, 1999). Thus, rates of obesity-related Type 2 diabetes are particularly high in Native Americans and Hispanics, and similar patterns have emerged in Australian Aborigines and the Indonesian Islanders (Zimmet, 1999). In children of South Asian descent, the move from rural to urban environments or migration to the U.S., U.K., and other European countries has been associated with a greatly increased prevalence of obesity-related disease and, in particular, Type 2 diabetes (McKeigue et al., 1991).

Genetic factors could underlie these population differences in risks for obesity-related disease. In particular, there may be common genetic polymorphisms that conferred some survival advantage during earlier times of undernutrition and through process of selection are now overrepresented in certain populations who have adapted to conditions of poor or intermittent nutrient supply. The original thrifty genotype hypothesis and subsequent debate has been centered on how genetic variations might enhance survival during adult life (Reaven, 1998). We would suggest that, in view of the relatively high mortality rates during perinatal life (Karn and Penrose, 1951) and in particular during times of nutritional debilitation (Ceesay et al., 1997), thrifty genotypes that evolved to enhance early perinatal survival may have a larger effect on reproductive fitness and selection advantage than genotypes that promote adult survival. Such "fetal thrifty genotypes" could now underlie current links between birth weight and adult disease risk.

The mean birth weight in any population is always slightly less than the optimal for perinatal survival of the infant (Alberman, 1991), and, as Moore and Haig pointed out, there is a complex paradigm where the interests of the mother may conflict with those of the father over fetal growth rates (Moore and Haig, 1991). Whereas it is in the interests of the father to have a larger baby, which achieves higher rates of perinatal survival and transmission of father's genes to subsequent generations, fetal overgrowth may be dangerous to the mother by making greater nutrient demands and by resulting in prolonged or obstructed labor. These conflicting interests of the mother and the father may have underpinned the evolution of genetic imprinting, a mechanism whereby only those genes transmitted from either the mother or father are expressed and the others are silenced (Reik and Walter, 2001b). A large proportion of those genes that are known to be imprinted are involved in the regulation of fetal growth; in general, genes that are exclusively paternally expressed enhance fetal growth, whereas exclusively maternally expressed genes are associated with reduced fetal growth (Reik and Walter, 2001a; Moore and Haig, 1991).

In infants who are smaller at birth, such as first-born infants who tend to be relatively restrained in utero, birth weight is more closely related to the birth weight of the mother and of the offspring of the mother's female relatives, suggesting a specific maternal line transmission of genetic factors that restrain fetal growth (Ounsted et al., 1986). We have reported genetic associations between size at birth and the mitochondrial DNA 16189 variant (Casteels et al., 1999), which is maternally inherited, and also with a common polymorphism in the maternally expressed *H19* gene (Ong et al., 2001). *H19* is a regulator of imprinting of the insulin-like growth factor-2 gene (*IGF2*), and its deletion in mice results in *IGF2* overexpression and larger birth size (Leighton et al., 1995). In the ALSPAC birth cohort, we observed that, particularly among restrained first pregnancies, a common *H19* genotype in the mother or offspring was associated with higher cord blood *IGF2* protein levels and larger birth size (Ong et al., 2001).

In average and larger birth weight babies, such as second- and third-born infants where fetal growth restraint is less evident, a more Mendelian pattern of birth weight inheritance is observed (Ounsted et al., 1986). In these pregnancies, a "variable number of tandem repeat" (VNTR) polymorphism adjacent to the insulin gene (*INS*), which regulates both *INS* and *IGF2* expression (Bennett et al., 1996; Paquette et al., 1998), has been repeatedly associated with size at birth, in particularly with head circumference at birth and also with *IGF2* protein levels in cord blood at birth (Dunger et al., 1998; Ong et al., 2004a). Although *IGF2* is paternally expressed, the birth size association with *INS* VNTR showed no parent of origin effect. Thus, with regard to human pregnancy, there is some evidence that maternal genes may restrain fetal growth; however, the role of exclusively paternally expressed genes in enhancing fetal growth has not yet been demonstrated.

Although we may be able to identify common genetic variations that are associated with size at birth, it is yet to be established whether these are thrifty or survival enhancing genotypes that could underpin future disease risk. The case is strongest for the *INS* VNTR, as this genetic polymorphism has been associated with postnatal weight gain, insulin resistance, central fat deposition, and Type 2 diabetes (Bennett and Todd, 1996; Waterworth et al., 1997; Ong et al., 1999b; Huxtable et al., 2000). One could argue that larger size at birth, conferred by the *INS* VNTR class III/III genotype, might improve fetal survival, but at the expense of increased subsequent risks for increased central adiposity, insulin resistance, and, ultimately, Type 2 diabetes. Other genes that promote the development of adiposity during early postnatal life could also have had selective advantages in promoting early survival. The development of central adiposity, although now linked with disease in today's overweight populations, could be considered as a survival advantage when nutrition was poor, as this central fat store provides a much more readily accessible source of nutrient during prolonged fasting or starvation compared with other body fat depots (Frayn, 1999). These hypotheses relating candidate thrifty genotypes to perinatal survival and long-term disease risks could be studied in African populations where there is considerable genetic diversity (Stead and Jeffreys, 2002) and particularly in some rural populations where variable nutrient supply still has an important impact on survival (Ceesay et al., 1997).

INTERACTION BETWEEN PHENOTYPE, GENOTYPE, AND OBESITY RISK

Experimental studies exploring the thrifty phenotype hypothesis in animal models have indicated that long-term obesity and disease risk markers can indeed be programmed by alterations in maternal nutrition such as protein restriction (Ozanne et al., 2004; Petry et al., 2001) or by reduced nutritional supply to the fetus by uterine artery ligation in late pregnancy (Simmons et al., 2001). However, the corollary of these observations that improving maternal nutrition may prevent future disease risk has not been supported by studies in humans. In contemporary populations, no clear relationship is seen between decreased maternal food intake and smaller size at birth (Rogers et al., 1998; Godfrey et al., 1997; Mathews et al., 1999), yet 25% of all infants still show fetal growth restriction and postnatal catch-up growth, which are known to be risk factors for the development of obesity and insulin resistance. In rural African populations where there is extremely poor maternal nutrition, maternal restraint of fetal growth may be further exaggerated and could contribute to increased long-term disease risks (Prentice et al., 1983). However, even in such populations, although improved maternal nutrition may benefit perinatal survival and reduce disease burden associated with poor nutrition in postnatal life (Ceesay et al., 1997), improved nutrition is unlikely to obviate long-term disease risk associated with obesity, such as Type 2 diabetes (van der Sande et al., 2001).

Studies in South Asian populations and most recently in East Germany, where there have been dramatic recent improvements in health and nutrition, show that the first effect of increased maternal nutrition is on greater maternal pregnancy weight gain (Hesse et al., 2003). Maternal weight gain is itself associated with increased risk for gestational diabetes, which for reasons yet to be fully elucidated also constitutes a risk factor for Type 2 diabetes in the offspring (Dabelea et al., 2000). Thus, against a genetic background that predisposes to obesity and gestational diabetes, the shift from poor to improved maternal nutrition could accelerate Type 2 diabetes risk rather than reduce it.

The mechanisms whereby gestational diabetes increases the risk for Type 2 diabetes in the offspring are yet not clear. One factor could be an effect of glucose toxicity on impaired insulin secretory response in the adult offspring (Sobngwi et al., 2003). A further explanation could lie in the association between Type 2 diabetes, gestational diabetes, and polycystic ovary syndrome (PCOS) (The Rotterdam-ESHRE/ASRM-Sponsored PCOS Consensus Workshop Group, 2004). Intriguingly, the polycystic ovary may improve fertility when nutrition is poor, yet it is associated with reduced fertility if obesity and insulin resistance should develop (Holte, 1998). Thus, it is not surprising that, in addition to obesity and Type 2 diabetes, the prevalence of PCOS is also increased in populations where there has been, until recently, a relatively poor nutrition (Rodin et al., 1998). In U.K. populations, PCOS is associated with larger birth weight (Michelmore et al., 1999; Cresswell et al., 1997), perhaps reflecting the effects of gestational diabetes or even high-normal glucose levels in the mother.

PCOS is associated with increased androgen production, and this in turn will lead to increased central adiposity (Ibanez et al., 2003b). Genetic factors associated with PCOS include the *INS* VNTR class III/III genotype (Waterworth et al., 1997) and shorter CAG repeat length in the androgen receptor gene, which confers greater receptor sensitivity (Ibanez et al., 2003c). These genetic factors increase androgen production and thus may themselves be seen as risk factors for increased central adiposity. Thus, in populations with a long history of undernutrition and a possible background genetic predisposition to PCOS, Type 2 diabetes, and gestational diabetes, a sudden change to abundant nutrition and obesity could rapidly lead to a vicious cycle of increasing childhood weight gain, increased risk of ovarian hyperandrogenism, central adiposity, and gestational diabetes, which in turn would increase the risk of Type 2 diabetes in the next generation.

RELEVANCE OF PERINATAL FACTORS TO THE CURRENT INCREASE IN CHILDHOOD OBESITY

The causes of the increasing prevalence of obesity during childhood are complex and are discussed in other chapters in this book. However, it should be remembered that the disease risks associated with obesity are not uniform (WHO Expert Consultation, 2004); i.e., not all obese children will have a similarly increased risk for cardiovascular disease and Type 2 diabetes. Although the numbers of obese children who show impaired glucose tolerance or other risk markers for cardiovascular disease during adolescence may be alarming (Sinha et al., 2002), there are still a large number of overweight or obese children who appear to have a low risk for the development of these diseases, at least in the short term. Understanding and more specific prediction in each individual of the disease risks associated with obesity could result from studies of interaction between early environmental and genetic factors (Ong et al., 2004a). In the presence of abundant nutrition and reduced energy expenditure, such interactions could determine the site of fat deposition and the associated development of insulin resistance. Furthermore, whereas insulin resistance may be a major risk factor for cardiovascular disease and Type 2 diabetes (Facchini et al., 2001), the ultimate development of Type 2 diabetes is determined by the ability of the cell to sustain compensatory hyperinsulinemia (Kahn, 2001). In recent studies in the ALSPAC birth cohort, we showed that early weight gain was the major determinant of insulin resistance; however, insulin secretion was more closely related to early height gain (Ong et al., 2004b). Thus, an understanding of early growth patterns and their associations with central adiposity and insulin resistance, and how these in turn may be modified by the genetic background, will be important in developing appropriate targeted interventions to prevent disease risk associated with obesity during childhood. Studies that combine early growth patterns and genetic factors to predict obesity and disease risks may also provide an opportunity to apply such targeted interventions at a much earlier stage.

REFERENCES

Alberman, E., Are our babies becoming bigger?, *J. R. Soc. Med.*, 84, 257–260, 1991.

Barker, D.J., Winter, P.D., Osmond, C., Margetts, B., and Simmonds, S.J., Weight in infancy and death from ischaemic heart disease, *Lancet,* 2, 577–580, 1989.

Bavdekar, A., Yajnik, C.S., Fall, C.H., Bapat, S., Pandit, A.N., Deshpande, V., Bhave, S., Kellingray, S.D., and Joglekar, C., Insulin resistance syndrome in 8-year-old Indian children: small at birth, big at 8 years, or both?, *Diabetes,* 48, 2422–2429, 1999.

Bennett, S.T. and Todd, J.A., Human type 1 diabetes and the insulin gene: principles of mapping polygenes, *Annu. Rev. Genet.,* 30, 343–370, 1996.

Bennett, S.T., Wilson, A.J., Cucca, F., Nerup, J., Pociot, F., McKinney, P.A., Barnett, A.H., Bain, S.C., and Todd, J.A., IDDM2-VNTR-encoded susceptibility to type 1 diabetes: dominant protection and parental transmission of alleles of the insulin gene-linked minisatellite locus, *J. Autoimmun.,* 9, 415–421, 1996.

Bundred, P., Kitchiner, P. and Buchan, I., Prevalence of overweight and obese children between 1989 and 1998: population based series of cross sectional studies, *BMJ,* 322, 326–328, 2001.

Casteels, K., Ong, K.K., Phillips, D.I., Bednarz, A., Bendall, H., Woods, K.A., Sherriff, A., Team, t.A.S., Golding, J., Pembrey, M.E., Poulton, J., and Dunger, D.B., Mitochondrial 16189 variant, thinness at birth and type 2 diabetes, *Lancet,* 353, 1499–1500, 1999.

Ceesay, S.M., Prentice, A.M., Cole, T.J., Foord, F., Weaver, L.T., Poskitt, E.M. and Whitehead, R.G., Effects on birth weight and perinatal mortality of maternal dietary supplements in rural Gambia: 5 year randomised controlled trial, *BMJ,* 315, 786–790, 1997.

Cresswell, J.L., Barker, D.J., Osmond, C., Egger, P., Phillips, D.I., and Fraser, R.B., Fetal growth, length of gestation, and polycystic ovaries in adult life, *Lancet,* 350, 1131–1135, 1997.

Dabelea, D., Hanson, R.L., Lindsay, R.S., Pettitt, D.J., Imperatore, G., Gabir, M.M., Roumain, J., Bennett, P.H. and Knowler, W.C., Intrauterine exposure to diabetes conveys risks for type 2 diabetes and obesity: a study of discordant sibships, *Diabetes,* 49, 2208–2211, 2000.

Dunger, D.B., Ong, K.K., Huxtable, S.J., Sherriff, A., Woods, K.A., Ahmed, M.L., Golding, J., Pembrey, M.E., Ring, S., Bennett, S.T., and Todd, J.A., Association of the INS VNTR with size at birth., *Nat. Genet.,* 19, 98–100, 1998.

Eriksson, J.G., Forsen, T., Tuomilehto, J., Winter, P.D., Osmond, C., and Barker, D.J., Catch-up growth in childhood and death from coronary heart disease: longitudinal study, *BMJ,* 318, 427–431, 1999.

Eriksson, J.G., Forsen, T.J., Osmond, C., and Barker, D.J., Pathways of infant and childhood growth that lead to type 2 diabetes, *Diabetes Care,* 26, 3006–10, 2003.

Facchini, F.S., Hua, N., Abbasi, F., and Reaven, G.M., Insulin resistance as a predictor of age-related diseases, *J. Clin. Endocrinol. Metabol.,* 86, 3574–3578, 2001.

Forsen, T., Eriksson, J.G., Tuomilehto, J., Teramo, K., Osmond, C., and Barker, D.J., Mother's weight in pregnancy and coronary heart disease in a cohort of Finnish men: follow up study, *BMJ,* 315, 837–840, 1997.

Frayn, K.N., Macronutrient metabolism of adipose tissue at rest and during exercise: a methodological viewpoint, *Proc. Nutr. Soc.,* 58, 877–886, 1999.

Gale, C.R., Martyn, C.N., Kellingray, S.D., Eastell, R., and Cooper, C., Intrauterine programming of adult body composition, *J. Clin. Endocrinol. Metabol.*, 86, 267–272, 2001.

Garnett, S.P., Cowell, C.T., Baur, L.A., Fay, R.A., Lee, J., Coakley, J., Peat, J.K., and Boulton, T.J., Abdominal fat and birth size in healthy prepubertal children, *Int. J. Obes. Relat. Metab. Disord.*, 25, 1667–1673, 2001.

Godfrey, K.M., Barker, D.J., Robinson, S., and Osmond, C., Maternal birthweight and diet in pregnancy in relation to the infant's thinness at birth, *Br. J. Obstetr. Gynaecol.*, 104, 663–667, 1997.

Golding, J., Pembrey, M.E., and Jones, R., ALSPAC—the Avon Longitudinal Study of Parents and Children. I. Study methodology, *Paediatr. Perinat. Epidemiol.*, 15, 74–87, 2001.

Hales, C.N., Barker, D.J., Clark, P.M., Cox, L.J., Fall, C., Osmond, C. and Winter, P.D., Fetal and infant growth and impaired glucose tolerance at age 64, *BMJ*, 303, 1019–1022, 1991.

Hales, C.N. and Barker, D.J., The thrifty phenotype hypothesis, *Br. Med. Bull.*, 60, 5–20, 2001.

Hediger, M.L., Overpeck, M.D., Kuczmarski, R.J., McGlynn, A., Maurer, K.R., and Davis, W.W., Muscularity and fatness of infants and young children born small- or large-for-gestational-age, *Pediatrics*, 102, E60, 1998.

Hesse, V., Voigt, M., Salzler, A., Steinberg, S., Friese, K., Keller, E., Gausche, R., and Eisele, R., Alterations in height, weight, and body mass index of newborns, children, and young adults in eastern Germany after German reunification, *J. Pediatr.*, 142, 259–262, 2003.

Holte, J., Polycystic ovary syndrome and insulin resistance: thrifty genes struggling with over-feeding and sedentary life style?, *J. Endocrinol. Invest.*, 21, 589–601, 1998.

Huxtable, S.J., Saker, P.J., Haddad, L., Walker, M., Frayling, T.M., Levy, J.C., Hitman, G.A., O'Rahilly, S., Hattersley, A.T., and McCarthy, M.I., Analysis of parent-offspring trios provides evidence for linkage and association between the insulin gene and type 2 diabetes mediated exclusively through paternally transmitted class III variable number tandem repeat alleles, *Diabetes*, 49, 126–130, 2000.

Ibanez, L., Ong, K., de Zegher, F., Marcos, M.V., del Rio, L., and Dunger, D.B., Fat distribution in non-obese girls with and without precocious pubarche: central adiposity related to insulinaemia and androgenaemia from prepuberty to post-menarche, *Clin. Endocrinol. (Oxf.)*, 58, 372–379, 2003a.

Ibanez, L., Ong, K.K., Ferrer, A., Amin, R., Dunger, D.B., and de Zegher, F., Low-dose flutamide-metformin therapy reverses insulin resistance and reduces fat mass in nonobese adolescents with ovarian hyperandrogenism, *J. Clin. Endocrinol. Metab.*, 88, 2600–2606, 2003b.

Ibanez, L., Ong, K.K., Mongan, N., Jaaskelainne, J., Marcos, M.V., Hughes, I.A., De Zegher, F., and Dunger, D.B., Androgen receptor gene CAG repeat polymorphism in the development of ovarian hyperandrogenism, *J. Clin. Endocrinol. Metab.*, 3333–3338, 2003c.

I Kahn, S.E., Clinical review 135: the importance of beta-cell failure in the development and progression of type 2 diabetes, *J. Clin. Endocrinol. Metab.*, 86, 4047–4058, 2001.

Karn, M.N. and Penrose, L.S., 1951 Birth weight and gestation time in relation to maternal age, parity and infant survival, *Ann. Eugen.*, 16, 147–158, 1951.

Law, C.M., Barker, D.J., Osmond, C., Fall, C.H., and Simmonds, S.J., Early growth and abdominal fatness in adult life, *J. Epidemiol. Community Health,* 46, 184–186, 1992.

Leighton, P.A., Ingram, R.S., Eggenschwiler, J., Efstratiadis, A., and Tilghman, S.M., Disruption of imprinting caused by deletion of the H19 gene region in mice, *Nature,* 375, 34–39, 1995.

Leon, D.A., Lithell, H.O., Vagero, D., Koupilova, I., Mohsen, R., and Berglund, L., Reduced fetal growth rate and increased risk of death from ischaemic heart disease: cohort study of 15000 Swedish men and women born 1915–29, *BMJ,* 317, 241–244, 1998.

Lindsay, R.S., Dabelea, D., Roumain, J., Hanson, R.L., Bennett, P.H., and Knowler, W.C., Type 2 diabetes and low birth weight. The role of paternal inheritance in the association of low birth weight and diabetes, *Diabetes,* 49, 445–449, 2000.

Mathews, F., Yudkin, P. and Neil, A., Influence of maternal nutrition on outcome of pregnancy: prospective cohort study, *BMJ,* 319, 339–343, 1999.

McCance, D.R., Pettitt, D.J., Hanson, R.L., Jacobsson, L.T., Knowler, W.C., and Bennett, P.H., Birth weight and non-insulin dependent diabetes: thrifty genotype, thrifty phenotype, or surviving small baby genotype?, *BMJ,* 308, 942–945, 1994.

McKeigue, P.M., Shah, B., and Marmot, M.G., Relation of central obesity and insulin resistance with high diabetes prevalence and cardiovascular risk in South Asians, *Lancet,* 337, 382–6, 1991.

McKeigue, P.M., Lithell, H.O., and Leon, D.A., Glucose tolerance and resistance to insulin-stimulated glucose uptake in men aged 70 years in relation to size at birth, *Diabetologia,* 41, 1133–1138, 1998.

Michelmore, K.F., Balen, A.H., Dunger, D.B., and Vessey, M.P., Polycystic ovaries and associated clinical and biochemical features in young women, *Clin. Endocrinol. (Oxf.),* 51, 779–786, 1999.

Moore, T. and Haig, D., Genomic imprinting in mammalian development: a parental tug-of- war, *Trends Genet.,* 7, 45–49, 1991.

Neel, J.V., Diabetes mellitus: a thrifty genotype rendered detrimental by "progress"?, *Am. J. Human Genet.,* 14, 353–362, 1962.

Ong, K.K., Ahmed, M.L., Sherriff, A., Woods, K.A., Watts, A., Golding, J., the-ALSPAC-Study-Team, and Dunger, D.B., Cord blood leptin is associated with size at birth and predicts infancy weight gain in humans, *J. Clin. Endocrinol. Metab.,* 84, 1145–1148, 1999a.

Ong, K.K., Phillips, D.I., Fall, C., Poulton, J., Bennett, S.T., Golding, J., Todd, J.A., and Dunger, D.B., The insulin gene VNTR, type 2 diabetes and birth weight, *Nat. Genet.,* 21, 262–263, 1999b.

Ong, K.K., Ahmed, M.L., Emmett, P.M., Preece, M.A., Dunger, D.B., and the-ALSPAC-Study-Team, Association between postnatal catch-up growth and obesity in childhood: prospective cohort study, *BMJ,* 320, 967–971, 2000.

Ong, K.K., Barratt, B., Kratzsch, J., Kiess, W., Pembrey, M.E., Team, t.A.S., Todd, J.A., and Dunger, D.B., Associations between cord blood IGF-II levels and common polymorphisms at INS VNTR and H19—genetic determinants of size at birth in humans, *Pediatr. Res.,* 49, 18A, 2001.

Ong, K.K., Preece, M.A., Emmett, P.M., Ahmed, M.L., and Dunger, D.B., Size at birth and early childhood growth in relation to maternal smoking, parity and infant breast-feeding: longitudinal birth cohort study and analysis, *Pediatr. Res.,* 52, 863–867, 2002.

Ong, K.K., Petry, C.J., Barratt, B.J., Ring, S., Cordell, H.J., Wingate, D.L., Pembrey, M.E., Todd, J.A., and Dunger, D.B., Maternal-fetal interactions and birth order influence insulin variable number of tandem repeats allele class associations with head size at birth and childhood weight gain, *Diabetes,* 53, 1128–1133, 2004a.

Ong, K.K., Petry, C.J., Emmett, P.M., Sandhu, M.S., Kiess, W., Hales, C.N., Ness, A.R., the-ALSPAC-study-team, and Dunger, D.B., Insulin sensitivity and secretion in normal children related to size at birth, postnatal growth, and plasma insulin-like growth factor-I levels., *Diabetologia,* 47, 1064–1070, 2004b.

Ounsted, M. and Sleigh, G., The infant's self-regulation of food intake and weight gain. Difference in metabolic balance after growth constraint or acceleration in utero, *Lancet,* 1, 1393–1397, 1975.

Ounsted, M., Scott, A., and Ounsted, C., Transmission through the female line of a mechanism constraining human fetal growth, *Ann. Human Biol.,* 13, 143–151, 1986.

Ozanne, S.E., Lewis, R., Jennings, B.J., and Hales, C.N., Early programming of weight gain in mice prevents the induction of obesity by a highly palatable diet, *Clin. Sci. (Lond.),* 106, 141–5, 2004.

Paquette, J., Giannoukakis, N., Polychronakos, C., Vafiadis, P., and Deal, C., The INS 5' variable number of tandem repeats is associated with IGF2 expression in humans, *J. Biol. Chem.,* 273, 14158–14164, 1998.

Petry, C.J., Dorling, M.W., Pawlak, D.B., Ozanne, S.E., and Hales, C.N., Diabetes in old male offspring of rat dams fed a reduced protein diet, *Int. J. Exp. Diabetes Res.,* 2, 139–143, 2001.

Prentice, A.M., Whitehead, R.G., Watkinson, M., Lamb, W.H., and Cole, T.J., Prenatal dietary supplementation of African women and birth-weight, *Lancet,* 1, 489–492, 1983.

Ravelli, G.P., Stein, Z.A., and Susser, M.W., Obesity in young men after famine exposure in utero and early infancy, *N. Engl. J. Med.,* 295, 349–353, 1976.

Reaven, G.M., Hypothesis: muscle insulin resistance is the ("not-so") thrifty genotype, *Diabetologia,* 41, 482–484, 1998.

Reik, W. and Walter, J., Evolution of imprinting mechanisms: the battle of the sexes begins in the zygote, *Nat. Genet.,* 27, 255–256, 2001a.

Reik, W. and Walter, J., Genomic imprinting: parental influence on the genome, *Nat. Rev. Genet.,* 2, 21–32, 2001b.

Reilly, J.J., Dorosty, A.R., and Emmett, P.M., Prevalence of overweight and obesity in British children: cohort study, *BMJ,* 319, 1039, 1999.

Rodin, D.A., Bano, G., Bland, J.M., Taylor, K., and Nussey, S. S., Polycystic ovaries and associated metabolic abnormalities in Indian subcontinent Asian women, *Clin. Endocrinol. (Oxf.),* 49, 91–99, 1998.

Rogers, I., Emmett, P., Baker, D., and Golding, J., Financial difficulties, smoking habits, composition of the diet and birthweight in a population of pregnant women in the South West of England. ALSPAC Study Team. Avon Longitudinal Study of Pregnancy and Childhood, *Eur. J. Clin. Nutr.,* 52, 251–260, 1998.

Simmons, R.A., Templeton, L.J., and Gertz, S.J., Intrauterine growth retardation leads to the development of type 2 diabetes in the rat, *Diabetes,* 50, 2279–2286, 2001.

Sinha, R., Fisch, G., Teague, B., Tamborlane, W.V., Banyas, B., Allen, K., Savoye, M., Rieger, V., Taksali, S., Barbetta, G., Sherwin, R.S., and Caprio, S., Prevalence of impaired glucose tolerance among children and adolescents with marked obesity, *N. Engl. J. Med.,* 346, 802–10, 2002.

Sobngwi, E., Boudou, P., Mauvais-Jarvis, F., Leblanc, H., Velho, G., Vexiau, P., Porcher, R., Hadjadj, S., Pratley, R., Tataranni, P.A., Calvo, F., and Gautier, J.F., Effect of a diabetic environment in utero on predisposition to type 2 diabetes, *Lancet,* 361, 1861–1865, 2003.

Soto, N., Bazaes, R.A., Pena, V., Salazar, T., Avila, A., Iniguez, G., Ong, K.K., Dunger, D.B., and Mericq, M.V., Insulin sensitivity and secretion are related to catch-up growth in small-for-gestational-age infants at age 1 year: results from a prospective cohort, *J. Clin. Endocrinol. Metab.,* 88, 3645–50, 2003.

Stead, J.D. and Jeffreys, A.J., Structural analysis of insulin minisatellite alleles reveals unusually large differences in diversity between Africans and non-Africans, *Am. J. Human Genet.,* 71, 1273–1284, 2002.

Stein, C.E., Fall, C.H., Kumaran, K., Osmond, C., Cox, V., and Barker, D.J., Fetal growth and coronary heart disease in south India, *Lancet,* 348, 1269–1273, 1996.

Stettler, N., Bovet, P., Shamlaye, H., Zemel, B.S., Stallings, V.A., and Paccaud, F., Prevalence and risk factors for overweight and obesity in children from Seychelles, a country in rapid transition: the importance of early growth, *Int. J. Obes. Relat. Metab. Disord.,* 26, 214–219, 2002a.

Stettler, N., Zemel, B.S., Kumanyika, S., and Stallings, V.A., Infant weight gain and childhood overweight status in a multicenter cohort study, *Pediatrics,* 109, 194–199, 2002b.

Tanner, J.M., *Human Growth; A Comprehensive Treatise,* Falkner, F., Ed., Plenum Press, New York, 1986, pp. 167–179.

The Rotterdam-ESHRE/ASRM-Sponsored PCOS Consensus Workshop Group, Revised 2003 consensus on diagnostic criteria and long-term health risks related to polycystic ovary syndrome (PCOS), *Human Reprod.,* 19, 41–47, 2004.

van der Sande, M.A., Ceesay, S.M., Milligan, P.J., Nyan, O.A., Banya, W.A., Prentice, A., McAdam, K.P., and Walraven, G.E., Obesity and undernutrition and cardiovascular risk factors in rural and urban Gambian communities, *Am. J. Public Health,* 91, 1641–1644, 2001.

Waterworth, D.M., Bennett, S.T., Gharani, N., McCarthy, M.I., Hague, S., Batty, S., Conway, G.S., White, D., Todd, J.A., Franks, S., and Williamson, R., Linkage and association of insulin gene VNTR regulatory polymorphism with polycystic ovary syndrome, *Lancet,* 349, 986–990, 1997.

WHO Expert Consultation, Appropriate body-mass index for Asian populations and its implications for policy and intervention strategies, *Lancet,* 363, 157–163, 2004.

Yajnik, C.S., Interactions of perturbations in intrauterine growth and growth during childhood on the risk of adult-onset disease, *Proc. Nutr. Soc.,* 59, 257–265, 2000.

Zimmet, P.Z., Diabetes epidemiology as a tool to trigger diabetes research and care, *Diabetologia,* 42, 499–518, 1999.

12

CHILDHOOD PREDICTORS OF BMI TRAJECTORIES

Tessa Parsons

CONTENTS

INTRODUCTION

There is now a growing body of literature describing the relationships between potential risk or protective factors and obesity, or body fatness. In both adults and children, there is more evidence for some risk factors than others. For example, there is consistent evidence that lower social position early in life increases the risk of obesity in adulthood and that children of fat parents are more likely to be fat themselves (Parsons et al., 1999). In contrast, evidence from longitudinal studies that dietary intake in childhood is related to subsequent fatness is very limited (Parsons et al., 1999). The literature on weight change and what influences change in weight, whether this is gain, loss, or maintenance, is largely restricted to adults (Fogelholm et al., 2000; Lahmann et al., 2000; Saris et al., 2003; Williamson, 1996), probably because of logistical study design issues and conceptual difficulties around weight change. Epidemiological studies of weight change will share the challenges of obesity studies; large numbers of participants and long-term follow-up are required, and there are inherent problems in measuring lifestyle predictors such as diet or physical activity. However, studies of weight change also require serial measurements of body weight. In terms of analyses, weight change is not necessarily a straightforward concept. The effect of a 90-kg person gaining 5 kg is different from that of a 60-kg person, so it could be argued that initial weight and/or height should be adjusted for. Alternatively, it might be preferable to look at change in, or maintenance of, body-mass index (BMI). This is acceptable in adults, but in children we expect weight and BMI to increase as part of normal growth, making the concept of weight change or maintenance more complex. In adults and children, maintenance requires definition — how much fluctuation in weight or BMI can be defined as maintenance?

Given these problems, and in order to be able to look at changes from childhood through to adulthood, in this chapter, we use data from the 1958 British birth cohort to look at trajectories of BMI from age 7 (or 16) to 42 years and examine the influence of several factors on the BMI trajectory. Factors were selected on the basis of them occurring in early life or childhood and having previously been found to be related to obesity (in some, but not necessarily all, studies); these included social class, parental BMI, birth weight, infant feeding method, childhood BMI, and pubertal maturation (Parsons et al., 1999). Here, the effect of these factors on the *rate of increase* in BMI is investigated.

1958 BRITISH BIRTH COHORT

All children born in England, Scotland, and Wales in the week of the 3rd to 9th of March 1958 were included in the 1958 British birth cohort (Butler and Bonham 1963). From a target population of 17,733 births, information was obtained on 98%. Major follow-ups of surviving children were conducted at ages 7, 11, 16, 23, 33, and 42 years [*Introduction to the National*

Child Development Study (NCDS), 2002; Ferri, 1993]. At age 42 years, 11,419 subjects from a target sample of 16,460 provided at least some information. Sample attrition has resulted in a slight underrepresentation of those who are most disadvantaged, but the remaining sample is generally representative of the original sample (Ferri, 1993). For analyses concerning factors in early life (social class, parental BMI, birth weight, and breast-feeding), multiple births were excluded ($n = 446$).

Body Mass Index (BMI)

BMI (kg/m^2) was calculated from heights and weights. At 7, 11, and 16 years, heights (to the nearest inch) and weights (in underclothes, to the nearest pound) were measured by trained medical personnel. At 23 and 42 years, self-reports of weight and height were obtained. Because of discrepancies between measured height at 33 years and self-reported height at 42 years (17% showed a difference of >3 cm), BMI at 42 years was calculated using height at 33 years, except where unavailable, in which case height at 42 years was used ($n = 422$). At 33 years, height was measured to the nearest centimeter, and weight measured with indoor clothing, without shoes, to the nearest 0.1 kg. Data at ages 23 and 33 years have been checked to detect coding errors (Lake, 1998; Power and Moynihan, 1988). BMI values for women who were pregnant at 33 years ($n = 256$) were excluded. At 23 and 42 years, pregnant women were asked to report prepregnant weight.

Predictors and Potential Confounding Factors

Two groups of predictors were considered: (a) early life predictors (social class, parental BMI, birth weight and breast-feeding), where the outcome was a BMI trajectory from 7 to 42 years (six time points), and (b) childhood predictors (BMI at 7 and 11 years and pubertal stage), where the outcome was a BMI trajectory from 16 to 42 years (four time points). Significant relationships were investigated for confounding.

Social class was defined in terms of father's occupation, according to the U.K. 1951 General Registrar's classification. Four categories are used in these analyses: (a) classes I and II (professional and managerial), (b) III-NM (skilled nonmanual), (c) III-M (skilled manual), and (d) IV, V (semi-skilled, unskilled manual); also recorded were those having "no male head of household."

Parental body size was measured as maternal height without shoes; prepregnant weight was self-reported in categories of 1 stone, shortly after the birth of the cohort member. Heights and weights for the fathers were reported by the mother when the child was 11 years old. Height was reported to the nearest inch; weights were classified into 1 of 27 groups, ranging from 6 stone 4 pounds (39.9 kg) to 19 stone 10 pounds (125.2 kg). To estimate BMI, parents were assigned a weight equivalent to the midpoint of their weight group. Parental BMI was expressed in tertiles for presentation purposes; tertile cut-offs for mother's BMI were 21.0798 and 23.5597 kg/m² and for father's BMI 23.3672 and 25.6267 kg/m².

Birth weight was recorded in pounds and ounces by the midwives in charge of the delivery and was converted to kilograms. Birth weight was split into tertiles for analyses due to nonlinear relationships with BMI (Parsons et al., 2001); cut-offs for tertiles were 3.1752 and 3.6288 kg for boys and 3.0618 and 3.4587 kg for girls.

Infant feeding was reported when the subject was 7 years, in four categories: breast-fed wholly or partially for (a) >1 month, (b) <1 month, (c) not at all, and (d) unknown. The latter group was small (0.9%) and was excluded from analyses.

Mother's smoking after the fourth month of pregnancy was reported shortly after the cohort member's birth. The mother was categorized as a nonsmoker (<1 cigarette per day) or smoker (≥1 cigarette per day).

BMI in childhood was expressed in tertiles for analyses, due to nonlinear relationships with BMI at subsequent ages (see above for measurement details). Tertile cut-offs for BMI were 15.2628 and 16.3306 kg/m² in boys and 14.9573 and 16.3116 kg/m² in girls at 7 years and 16.1287 and 17.6232 kg/m² in boys and 16.2294 and 18.1694 kg/m² in girls at 11 years.

Puberty was assessed at 11 and 16 years using several ratings; two each for girls and boys are presented here. Doctors assessed breast development in girls at 11 years (on a scale of 5, with 1 = preadolescent and 5 = mature) and axillary hair pattern in boys at 16 years (absent, sparse, intermediate, or adult) (Tanner, 1962). For analyses, breast development was collapsed into three categories, stage 1, stage 2, and stages 3-5, and axillary hair pattern into two categories, more advanced (intermediate or adult pattern) and less advanced (absent or sparse). At the age 16 years medical examination, girls were asked at what age they had first menstruated; three categories were used in analyses: 9–11 years, 12–13 years, and 14–17 years. Also at 16 years, the parent reported what age the boy's voice had broken; four categories were used in analysis: ≤12, 13, 14, and ≥15 years.

STATISTICAL METHODS

Multilevel models, which allow for correlated data within individuals measured on several occasions, were fitted with BMI as a repeated outcome measure. These models allow for missing outcome data; i.e., all those with at least one BMI measure (and data on all independent variables) are included in the model. It is assumed that outcome data are missing at random, i.e., that the likelihood of a measurement being missing is unrelated to the actual measurement value. All analyses were performed separately for males and females, using computer package MLWin version 1.10, Institute of Education, London, U.K. (Rasbash et al., 2000). First, change in BMI with age was modeled, using both linear and quadratic age terms. Because each follow-up of the cohort members took place over a period of time (6–18 months), exact age (to the nearest month) was used wherever possible. The intercept of the model represents mean BMI at the first time point for the outcome variable, i.e., age 7 years for the early life models and age 16 years for the childhood models (age was centered at 7 years for the early life models and 16 years for the childhood models). The slope represents the linear change in BMI per year. Random effects at the individual level (level 2) were included for the intercept and linear age term, allowing both intercept and linear component of the slope to vary between individuals. Random effects for the quadratic age term were not significant. The model also included a random effect at level 1 (measurement occasion). Second, to look, for example, at the influence of social class on change in BMI, social class was added to the model, allowing the intercept (BMI at age 7) to vary by social class. A social class by age interaction term was also added, to test whether the slope of the BMI trajectory varied by social class. The effect of each predictor was investigated similarly and in turn, adjusting for potentially confounding factors and their interaction terms by age where appropriate.

FINDINGS

The numbers of cohort members with any information (not necessarily including BMI) at each follow-up are summarized in Table 12.1, together with potential predictors of the BMI trajectory investigated here, and possible confounding factors.

Table 12.1 Follow-Ups of 1958 British Birth Cohort and Variables Included in Analysis

		Age at Follow-Up (years)					
	Birth (1958)	7	11	16	23	33	42(1999 /2000)
N*	17,414	15,468	15,303	14,761	12,537	11,407	11,419
Variables		BMI	BMI	BMI	BMI	BMI	BMI
	SES	Breast feeding	Father's BMI	Puberty			
	Mother's BMI						
	Birth weight						

*N = number of cohort members with at least some information collected.

Table 12.2   BMI at Each Age

	Males		Females	
Age (years)	Mean (SD), kg/m²	N	Mean (SD), kg/m²	N
7	15.95 (1.63)	6717	15.89 (1.91)	6268
11	17.31 (2.41)	6233	17.66 (2.71)	5982
16	20.25 (2.73)	5577	21.02 (2.96)	5233
23	23.11 (2.91)	5999	22.13 (3.21)	6008
33	25.63 (4.01)	5375	24.60 (4.87)	5308
42	29.90 (4.65)	5380	28.53 (5.76)	5476

BMI increases with age; mean BMI at each age is shown in Table 12.2, and the BMI trajectory, modeled with age and age² terms, for males and females separately, is shown in Figure 12.1.

Early Life Factors (Outcome BMI 7 to 42 Years)

Social Class

Social class had no influence on BMI at age 7 years (intercept) in either males or females (Table 12.3). However, in both sexes, there was a significant effect on the slope of the trajectory, such that, in males from social classes IV

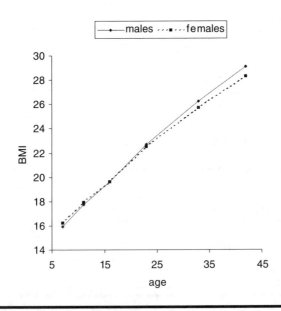

Figure 12.1   BMI Trajectories From 7 to 42 Years for Males and Females.

Table 12.3 Effect of early life factors on BMI between age 7 and age 42 years

	Males			Females		
	Intercept at age 7 Years (kg/m²) Coefficient (95% CI)	Linear Change from Age 7 Years (kg/m²/year) Coefficient (95% CI)	N	Intercept at Age 7 Years (kg/m²) Coefficient (95% CI)	Linear Change from Age 7y (kg/m²/year) Coefficient (95% CI)	N
Social class						
III-NM vs. I and II	-0.19 (-0.36 to 0.01)	0.015 (0.003 to 0.027)		-0.13 (-0.33 to 0.08)	0.028 (0.012 to 0.044)	
III-M vs. I and II	0.00 (-0.12 to 0.12)	0.033 (0.023 to 0.043)		-0.06 (-0.20 to 0.09)	0.049 (0.037 to 0.061)	
IV and V[a] vs. I and II	-0.06 (-0.20 to 0.08)	0.046 (0.036 to 0.056)	7718	-0.10 (-0.26 to 0.07)	0.059 (0.047 to 0.071)	7327
Mother's BMI						
Middle vs. bottom tertile	0.32 (0.21 to 0.43)	0.010 (0.002 to 0.018)		0.46 (0.34 to 0.59)	0.007 (-0.003 to 0.017)	
Top vs. bottom tertile	0.88 (0.77 to 0.99)	0.028 (0.020 to 0.036)	7465	1.12 (1.00 to 1.25)	0.040 (0.030 to 0.050)	7068
Father's BMI						
Middle vs. bottom tertile	0.30 (0.18 to 0.42)	0.016 (0.008 to 0.024)		0.36 (0.22 to 0.49)	0.009 (-0.001 to 0.019)	
Top vs. bottom tertile	0.70 (0.58 to 0.82)	0.038 (0.030 to 0.046)	6479	0.82 (0.69 to 0.95)	0.035 (0.025 to 0.045)	6129

(continued)

Table 12.3 Effect of Early Life Factors on BMI Between Age 7 and Age 42 Years (Continued)

	Males			Females		
	Intercept at Age 7 Years (kg/m²) Coefficient (95% CI)	Linear Change from Age 7Years (kg/m²/year) Coefficient (95% CI)	N	Intercept at Age 7 Years (kg/m²) Coefficient (95% CI)	Linear Change from Age 7y (kg/m²/year) Coefficient (95% CI)	N
Mother and father's BMI combined						
Combination[b] vs. both in bottom tertile	0.73 (0.57 to 0.88)	0.027 (0.015 to 0.039)		0.84 (0.67 to 1.02)	0.018 (0.004 to 0.032)	
Both in top vs. both in bottom tertile	1.65 (1.45 to 1.85)	0.062 (0.048 to 0.076)	5854	1.91 (1.68 to 2.13)	0.069 (0.051 to 0.087)	5547
Birth weight						
Middle vs. bottom tertile	0.24 (0.13 to 0.34)	−0.005 (0.013 to 0.003)		0.22 (0.09 to 0.34)	−0.012 (−0.022 to −0.002)	
Top vs. bottom tertile	0.63 (0.52 to 0.74)	0.002 (−0.006 to 0.010)	7633	0.63 (0.50 to 0.75)	−0.008 (−0.018 to 0.002)	7284
Adjusted for [c]SES, [d]m-BMI, [e]m-smoke						
Middle vs. bottom tertile	0.19 (0.08 to 0.30)	−0.003 (−0.011 to 0.005)		0.15 (0.02 to 0.28)	−0.011 (−0.021 to 0.001)	
Top vs. bottom tertile	0.51 (0.40 to 0.63)	0.002 (−0.006 to 0.010)	6958	0.47 (0.34 to 0.60)	−0.009 (−0.019 to 0.001)	6664

Breast-feeding							
Breast-fed <1 month vs. not breast-fed	−0.10	(−0.22 to 0.02)	0.001	(−0.009 to 0.011)	−0.06	(−0.20 to 0.08)	−0.008	(−0.020 to 0.004)
Breast-fed >1 month vs. notbreast-fed	−0.04	(−0.14 to 0.06)	−0.008	(−0.016 to 0.000) 7222	−0.04	(−0.16 to 0.08)	−0.015	(−0.025 to −0.005) 7222
Adjusted for [c]SES, [d]m-BMI, [e]m-smoke							
Breast-fed <1 month vs. not breast-fed	−0.03	(−0.16 to 0.09)	0.001	(−0.009 to 0.011)	−0.11	(−0.25 to 0.04)	−0.001	(−0.013 to 0.011)
Breast-fed >1 month vs. not breast-fed	0.04	(−0.07 to 0.15)	−0.003	(−0.011 to 0.005) 6396	−0.03	(−0.15 to 0.010)	−0.008	(−0.018 to 0.002) 6070

[a] Includes cohort members with no male head of household.
[b] Combination: parents in different tertiles or both mother and father in middle tertile.
[c] SES = social class.
[d] m-BMI = mother's BMI.
[e] m-smoke = mother's smoking during pregnancy.

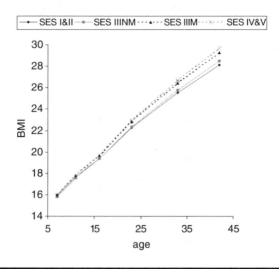

Figure 12.2   Male BMI Trajectories for Ages 7 to 42 Years by Social Class.

and V, BMI increased at a higher rate, 0.046 kg/m²/year, than in males from social classes I and II (Table 12.3, Figure 12.2). In females, this difference in slopes was greater, at 0.059 kg/m²/year (Table 12.3). Thus, although in both sexes there was no difference in BMI between the social classes at 7 years, at 42 years, males from social classes IV and V had BMIs 1.5 kg/m² higher and females 2.0 kg/m² higher than their counterparts in social classes I and II. In both sexes, the intermediate social classes, III-NM and III-M, had intermediate slopes, such that the slope of the BMI trajectory became steeper from social classes I and II to social classes IV and V. The trajectory slopes for social classes I and II were significantly less steep than all other groups (Table 12.3).

Parental BMI

Mother's BMI had a significant effect on both BMI at age 7 years (intercept) and the slope of the BMI trajectory (Table 12.3). Boys whose mother's BMI was in the top or medium tertile compared with the bottom tertile had higher BMIs at age 7, by 0.88 and 0.32 kg/m², respectively, and higher rates of BMI increase between age 7 and 42, by 0.010 kg/m²/year and 0.028 kg/m²/year, respectively (Table 12.3, Figure 12.3). Similar and slightly stronger effects were seen in females (Table 12.3). Father's BMI had much the same effect but was independent of mother's BMI. When mother's and father's BMIs were combined, effects on both BMI at age 7 years and the BMI slope were stronger; boys and girls with both parents in the top BMI tertile had higher BMIs at 7 years, by 1.65 and 1.91 kg/m², respectively,

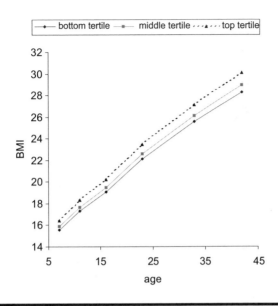

Figure 12.3   Male BMI Trajectories For Ages 7 to 42 Years by Tertile of Mother's BMI.

compared with children with both parents in the bottom BMI tertile. They also had faster rates of increase through to adulthood, by 0.062 kg/m²/year in males and 0.069 kg/m²/year in females (Table 12.3). Thus, by age 42 years, males with parents in the top BMI tertile had a BMI 3.8 kg/m² greater than males with parents in the bottom BMI tertile. In females, the difference was 4.3 kg/m². The effects of parental BMI and social class were independent (models not shown).

Birth Weight

Boys and girls with higher birth weights had higher BMIs at age 7 years; the difference between top and bottom tertile was 0.63 kg/m² in both sexes (Table 12.3). In males, birth weight had no influence on the slope of the BMI trajectory. In females, the slope of the BMI trajectory for the top tertile of birth weight did not differ significantly from the bottom tertile. However, those whose birth weight was in the middle tertile showed a *slower* BMI increase than those whose birth weight was in the bottom tertile, by 0.012 kg/m²/year, although this was only just significant (Table 12.3). Thus, although at 7 years middle birth weight females had a higher BMI than lower birth weight females, at 42 years they had a *lower* BMI. Adjusting for mother's BMI, social class, mother's smoking during pregnancy, and inter-action terms of each of these factors by age slightly reduced this effect (Table 12.3, Figure 12.4).

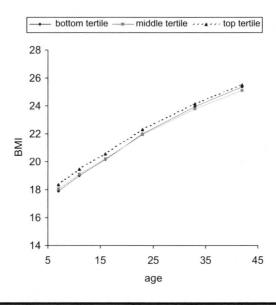

Figure 12.4   Female BMI Trajectories for Ages 7 to 42 Years by Tertile of Birth Weight Adjusted for Social Class, Mother's BMI, and Mother's Smoking in Pregnancy.

Breast-Feeding

Breast-feeding had no influence on BMI at age 7 years, but children who were breast-fed for more than 1 month had slower increases in BMI than children who were never breast-fed, by 0.008 kg/m²/year in males (only borderline significance) and by 0.015 kg/m²/year in females (Table 12.3). Adjusting for social class, mother's BMI, mother's smoking in pregnancy, and interaction terms of each of these factors by age reduced the slope differences, which were no longer significant (Table 12.3).

Childhood Factors (Outcome BMI 16 to 42 Years)

BMI at Age 7 and 11 Years

Both boys and girls who had a higher BMI at age 7 years had a higher BMI at age 16 years, but they also had significantly steeper rates of increase between 16 and 42 years (Table 12.4, Table 12.5, and Figure 12.5). The slope difference between the top and bottom tertiles was 0.015 kg/m²/year for males and 0.014 kg/m²/year for females. The effect of BMI at 11 years was stronger, especially in girls; those in the top tertile had steeper BMI trajectories than those in the bottom tertile, by 0.023 kg/m²/year in males and 0.051 kg/m²/year in females (Table 12.4 and Table 12.5).

Table 12.4 Effect of Childhood BMI and Puberty on BMI Between Age 16 and Age 42 Years in Males

	Intercept at Age 16 Years (kg/m²) Coefficient (95% CI)		Linear Change from Age 16 Years (kg/m²/year) Coefficient (95% CI)		N
BMI at 7 years					
Middle vs. bottom tertile	0.93	(0.77 to 1.09)	0.002	(−0.010 to 0.014)	
Top vs. bottom tertile	2.59	(2.43 to 2.75)	0.015	(0.003 to 0.027)	6121
BMI at 11 years					
Middle vs. bottom tertile	1.42	(1.27 to 1.56)	0.010	(0.000 to 0.020)	
Top vs. bottom tertile	3.59	(3.44 to 3.73)	0.023	(0.013 to 0.033)	5870
Axillary hair pattern at 16 years					
More advanced vs. less advanced	1.03	(0.89 to 1.17)	−0.014	(−0.024 to −0.004)	5602
Adjusted for BMI at 11 years					
More advanced vs. less advanced	0.43	(0.31 to 0.54)	−0.013	(−0.023 to −0.003)	4460
Age of voice breaking					
14 vs. ≥15 years	0.71	(0.54 to 0.88)	0.006	(−0.006 to 0.018)	
13 vs. ≥15 years	1.08	(0.87 to 1.29)	0.002	(−0.012 to 0.016)	
≤12 vs. ≥15 years	1.41	(1.16 to 1.66)	0.019	(0.003 to 0.035)	5386
Adjusted for BMI at 11 years					
14 vs. ≥15 years	0.38	(0.24 to 0.52)	0.004	(−0.008 to 0.016)	
13 vs. ≥15 years	0.61	(0.44 to 0.78)	0.003	(−0.013 to 0.019)	
≤12 vs. ≥15 years	0.75	(0.54 to 0.95)	0.014	(−0.004 to 0.032)	4296

Puberty

Boys with more advanced axillary hair pattern (intermediate or adult) had a higher BMI at 16 years, by 1.0 kg/m², but a slower rate of increase, by 0.014 kg/m²/year, than boys with a less advanced axillary hair pattern (absent or sparse) (Table 12.4). After we adjusted for BMI at 11 years, the BMI

Table 12.5 Effect of Childhood BMI and Puberty on BMI Between Age 16 and Age 42 Years in Females

	Intercept at Age 16 Years (kg/m²) Coefficient (95% CI)		Linear Change from Age 16 Years (kg/m²/year) Coefficient (95% CI)		N
BMI at 7 years					
Middle vs. bottom tertile	1.23	(1.06 to 1.40)	0.000	(−0.014 to 0.014)	
Top vs. bottom tertile	3.12	(2.95 to 3.29)	0.014	(0.000 to 0.028)	5814
BMI at 11 years					
Middle vs. bottom tertile	1.64	(1.49 to 1.80)	0.012	(−0.002 to 0.026)	
Top vs. bottom tertile	3.92	(3.76 to 4.07)	0.051	(0.037 to 0.065)	5674
Breast stage at 11 years[a]					
Stage 2 vs. 1	0.90	(0.72 to 1.07)	0.017	(0.003 to 0.031)	
Stage 3–5 vs. 1	1.78	(1.60 to 1.97)	0.023	(0.009 to 0.037)	5782
Adjusted for BMI at 11 years					
Stage 2 vs. 1	0.06	(−0.08 to 0.20)	0.008	(−0.006 to 0.022)	
Stage 3–5 vs. 1	−0.02	(−0.17 to 0.14)	0.005	(−0.009 to 0.019)	5647
Age at menarche					
12–13 vs. ≥14 years	0.99	(0.80 to 1.17)	0.012	(−0.002 to 0.026)	
9–11 vs. ≥14 years	1.96	(1.70 to 2.21)	0.060	(0.040 to 0.080)	4419
Adjusted for BMI at 11 years					
12–13 vs. ≥14 years	0.18	(0.02 to 0.34)	0.006	(−0.010 to 0.022)	
9–11 vs. ≥14 years	0.13	(−0.09 to 0.35)	0.042	(0.020 to 0.064)	3591

[a] Breast stage: 1 = preadolescent, 5 = mature.

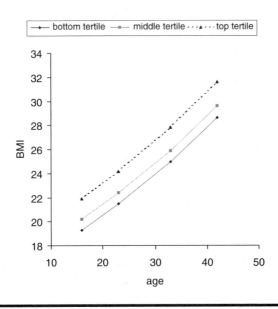

Figure 12.5 Male BMI Trajectories for Ages 16 to 42 Years by Tertile at 7 Years.

difference at 16 years was reduced to 0.4 kg/m²/year and remained signif-icant, and the difference in rate of increase remained the same — lower for boys with more advanced maturation. Boys whose voice broke earlier, ≤12 years compared with ≥15 years, had a higher BMI at 16 years, by 1.4 kg/m², but a higher rate of increase, by 0.019 kg/m²/year (Table 12.4, Figure 12.6A). After we adjusted for BMI at 11 years, the difference at 16 years was reduced, as was the slope difference, which was no longer significant (Table 12.4, Figure 12.6B).

Girls with more advanced breast development at 11 years had a higher BMI at 16 years, by 1.78 kg/m² for stages 3–5 compared with stage 1, and a faster rate of increase, by 0.023 kg/m²/year (Table 12.5). After we adjusted for BMI at 11 years, the differences at 16 years and in the rate of increase were abolished (Table 12.5). The earlier the age of menarche, the higher the BMI in girls at 16 years, and girls in the earliest menarche group had a steeper BMI trajectory than girls in the latest menarche group, by 0.06 kg/m²/year (Table 12.5, Figure 12.7A). After we adjusted for BMI at 11 years, the differences in BMI at 16 years were eliminated, but the slope difference between girls in the 9 to 11 year and 14 year or later groups remained significant, at 0.04 kg/m²/year (Table 12.5, Figure 12.7B).

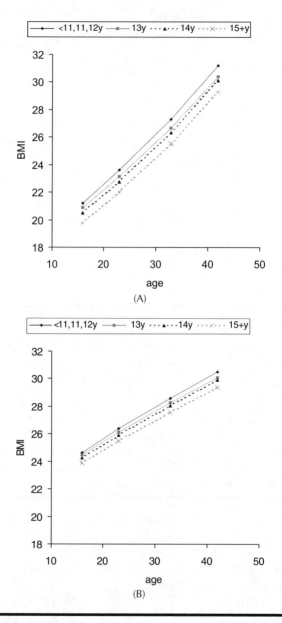

Figure 12.6   A: Male BMI Trajectories for Ages 16 to 42 Years by Age of Voice Breaking. B: Male BMI Trajectories for Ages 16 to 42 Years by Age of Voice Breaking Adjusted for BMI at 11 Years.

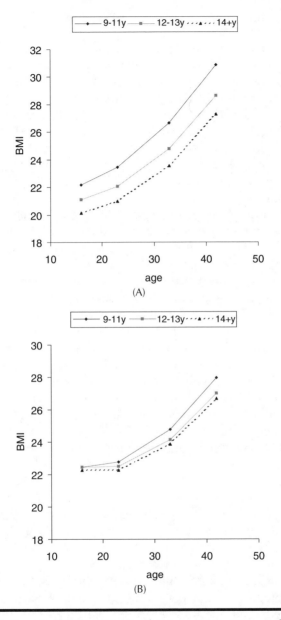

Figure 12.7 A: Female BMI Trajectories for Ages 16 to 42 Years by Age of Menarche. B: Female BMI Trajectories for Ages 16 to 42 Years by Age of Menarche Adjusted for BMI at 11 Years.

DISCUSSION

This study shows that some factors already recognized as predictors of obesity, namely, social class at birth, parental BMI, BMI in childhood, and, in girls, age of menarche, have an additional effect on the rate of increase in body mass through to 42 years. Thus the effect of these predictors becomes amplified over time. Previous analyses of obesity and BMI in the 1958 birth cohort have used logistic and linear regression techniques to examine relationships between predictors and body fatness. This chapter builds on the previous work, but we also use trajectory or "growth" models in a multi-level framework to provide additional information on change in BMI over time. Results from these multilevel models cannot be predicted from results from separate regression analyses, as the following examples illustrate. In separate regression analyses, the coefficients for the relationships between social class at birth and BMI at each time-point (7, 11, 16, 23, 33, and 42 years) increase over time (Power and Matthews, 1997), whereas the coefficients for the relationships between BMI at 7 years and each subsequent time point show a slight decrease over time (Power et al., 1997a). However, the multilevel models demonstrate that both social class at birth and BMI at 7 years affect change in BMI over time. Children born into a lower social class, or with a higher BMI at 7 years, are set on a steeper BMI trajectory than their respective peers born into a higher social class or with a lower BMI at 7 years.

Clearly, an advantage of longitudinal cohort data is that data at several time points are available, allowing this type of analysis. Another major advantage of the 1958 cohort is that it provides information on a wide range of potential confounding factors. There are some limitations to the data; mother's weight and father's weight and height were reported rather than measured, and maternal smoking and breast-feeding were reported retrospectively. However, these limitations have been shown to be unlikely to have a major impact on relations with BMI or obesity in the 1958 cohort (Lake et al., 1997; Li et al., 2003; Parsons et al., 2003; Power and Jefferis, 2002).

The effect of social class on the BMI trajectory in the 1958 birth cohort was not explained by mother's BMI, birth weight, breast-feeding, or mother's smoking in pregnancy. Even in combination, these factors only slightly reduced the effect of social class on the BMI trajectory. The relationship between lower social group at birth and higher levels of adult obesity (at a single time point) is remarkably consistent across industrialized countries (Parsons et al., 1999) and is similarly unexplained. Possible explanatory factors include dietary habits and physical activity, which require further investigation. Lifestyle habits tend to cluster, such that individuals from higher social groups engage in a number of more healthy lifestyle habits (Health Survey for England, 1999; Whichelow and Prevost, 1996), and social class may be capturing patterns of behavior. However, the fact that social class in early life has a stronger effect on subsequent fatness than social class in adolescence or adulthood suggests that, independent of adult education, there is something about circumstances in early life that has a persistent effect (Power et al., 2003). It is also possible that the influences of early life circumstances is dependent on environmental conditions and that it is the interaction between the two that is important. It is worth noting that eating and physical activity patterns have seen dramatic changes during the lives of the 1958 cohort members, with increasing motorized transport, reduced manual labor at home and work, and increasing high-fat diets (Prentice and Jebb, 1995).

Mother's and father's body size make a similar contribution to their offspring's BMI trajectory, and their effects are independent. Therefore, when both parents have a high BMI, their offspring has a higher rate of increase between age 7 and 42 years than when only one parent has a high BMI. The effect of parental BMI will be exerted both through inherited genes and shared lifestyles, and estimates as to which is most important vary (Parsons et al., 1999). The amplification of the effect seen with age in these analyses may be due to the cumulative effect of obesity-promoting lifestyle habits and/or an increasingly obesogenic environment allowing the expression of genes predisposing to obesity, which in conditions of food limitation or high physical demands would not be expressed. It is becoming evident that parental BMI is important not only because of its relationship with offspring obesity but because, as previously found in the 1958 cohort, children of obese parents show stronger child to adult BMI correlations (Lake et al., 1997), and, as seen here, children of fatter parents show higher rates of BMI increase through to adulthood.

Higher birth weight has been found by several studies to predict obesity (Parsons et al., 1999; Whitaker and Dietz, 1998), but this may be due to maternal body size and a reflection of larger mothers having larger babies (Parsons et al., 2001). In the current study, birth weight did not influence the change in BMI over time in males. In females, girls with a birth weight in the bottom tertile had a steeper BMI trajectory than girls whose birth weight was in the top tertile, so that lower birth weight girls appeared to "catch up" with the higher birth weight girls. Other studies have shown similar relationships between low size at birth and rapid weight gain or "catch-up growth" and subsequent fatness, but much earlier in life (Ong et al., 2000; Ong et al., 2002). The effect over a much longer time period in females in the 1958 birth cohort is intriguing, but its importance is uncertain since the size of the effect was small and of borderline significance.

The effect of breast-feeding on obesity remains inconclusive; several recent and large studies have shown a protective effect or no effect

(Gillman et al., 2001; Hediger et al., 2001; Li et al., 2003). Secular trends do not support a protective effect; in the U.K. since 1990, breast-feeding incidence has increased, and prevalence has remained stable (Department of Health, 2002), while obesity rates have soared (Health Survey for England, 1999; Rudolf et al., 2001). Although the multilevel analysis presented here showed that children breast-fed more than a month had slower rates of BMI increase than children not breast-fed, this effect disappeared after adjusting for social class, mother's BMI, and mother's smoking during pregnancy, as did the protective effect of breast-feeding on obesity at 33 years in earlier regression analyses (Parsons et al., 2003). Breast-feeding is important for many aspects of child health, but findings from the 1958 cohort suggest it is unlikely to have a significant role in obesity prevention.

It is well-known that obesity tends to track and that fat children are more likely to become fat adults. (Power et al., 1997b; Power et al., 1997a). In the present analyses, children who were in the top tertile of BMI at 7 years, instead of their BMI from age 16 to 42 years increasing in parallel with their peers, had a steeper trajectory, although significance in girls was borderline. BMI at 11 years showed similar but stronger effects, with the steeper trajectory particularly evident for girls in the top tertile. The persistent effect of childhood fatness on change in BMI over time has also been found in the U.K. 1946 birth cohort (Hardy et al., 2000) and in both the 1946 and 1958 cohorts was independent of social class. Again, the amplification effect could be due to the cumulative effects of lifestyle habits that promote obesity, such as low physical activity levels and high energy intake, and/or gene-environment interaction effects.

Many studies have shown a relationship between markers of puberty and subsequent fatness such that those who mature earlier or faster are fatter postmaturation (Parsons et al., 1999). However, there is also evidence that children who are fatter before puberty mature earlier, as shown in the 1958 cohort (Power et al., 1997a; Stark et al., 1989) and other studies (Biro et al., 2003; Freedman et al., 2003; Van Lenthe et al., 1996). When prepubertal fatness is accounted for, the relationship between age at puberty and later fatness is greatly reduced (Freedman et al., 2003; Power et al., 1997a). In the present study, the effects for different markers of puberty on the BMI trajectory were not entirely consistent. In boys, the earlier the voice broke, the higher the BMI at age 16 years and the steeper the BMI trajectory, although the slope differences disappeared after adjusting for BMI at age 11 years. Findings for age of menarche and breast stage at 11 years in girls were similar, although, for age of menarche, the steeper BMI trajectory for earlier maturing girls remained significant after adjusting for BMI at 11 years. In boys, even before adjusting for BMI at 11 years, the BMI trajectory was *less steep* in those with a more advanced pattern of axillary hair at 16 years and remained so after adjustment. Another study has suggested that, in girls, those who show breast development before pubic hair development have higher pre- and postpubertal BMI than girls who show pubic hair development before breast development, despite age of pubertal onset being the same (Biro et al., 2003). Different markers of puberty may therefore be providing information about different patterns of growth and body fat, and understanding of these interrelationships is currently limited.

In summary, several childhood factors increase the rate of gain in BMI from childhood through to adulthood, including lower social class at birth, high parental BMI, high BMI in childhood, and, in girls, earlier age of menarche. The mechanisms by which these factors exert their effects are not yet fully understood, but high parental and childhood BMI might provide a focus for management and prevention of future BMI gain.

ACKNOWLEDGMENTS

Data were obtained from the U.K. Data Archive, University of Essex (files: National Child Development Study, SN 3148, SN 4396). Data providers included Centre for Longitudinal Studies, Institute of Education and National Birthday Trust Fund, National Children's Bureau, and City University Social Statistics Research Unit (original data producers).

REFERENCES

Biro, F.M., Lucky, A.W., Simbartl, L.A., Barton, B.A., Daniels, S.R., Striegel-Moore R., Kronsberg S.S., and Morrison J.A., Pubertal maturation in girls and the relationship to anthropometric changes: pathways through puberty, *J. Pediatr.,* 142, 643–646, 2003.

Butler, N.R. and Bonham, D.G., *Perinatal mortality,* Edinburgh: Churchill Livingstone, 1963.

Department of Health, *Infant Feeding 2000,* available at http://www.dh.gov.uk/assetRoot/04/05/97/63/04059763.pdf, The Stationery Office, London, 2002.

Ferri, E., *Life at 33: The Fifth Follow-Up of the National Child Development Study,* National Children's Bureau, London, 1993.

Fogelholm, M. and Kukkonen-Harjula, K., Does physical activity prevent weight gain — a systematic review, *Obes. Rev.,* 1, 95–111, 2000.

Freedman, D.S., Khan, L.K., Serdula, M.K., Dietz, W.H., Srinivasan, S.R., and Berenson, G.S., The relation of menarcheal age to obesity in childhood and adulthood: the Bogalusa heart study, *BMC Pediatr.,* 3, 3, 2003.

Gillman, M.W., Rifas-Shiman, S.L., Camargo, C.A., Jr., Berkey, C.S., Frazier, A.L., Rockett HR., Field, A.E., and Colditz, G.A., Risk of overweight among adolescents who were breastfed as infants, *JAMA,* 285, 2461–2467, 2001.

Hardy, R., Wadsworth, M., and Kuh, D., The influence of childhood weight and socioeconomic status on change in adult body mass index in a British national birth cohort, *Int. J. Obes.,* 24, 725–734, 2000.

Health Survey for England, *Cardiovascular Disease '98,* The Stationery Office, London, 1999.

Hediger, M.L., Overpeck, M.D., Kuczmarski, R.J., and Ruan, W.J., Association between infant breastfeeding and overweight in young children. *JAMA*, 285, 2453–2460, 2001.

Introduction to the National Child Development Study (NCDS), available at http://www.cls.ioe.ac.uk/Cohort/Ncds/Documentation/maindocs.htm (Accessed 16 March 2004), 2002

Lahmann, P.H., Lissner, L., Gullberg, B., and Berglund, G., Sociodemographic factors associated with long-term weight gain, current body fatness and central adiposity in Swedish women, *Int. J. Obes.*, 24, 685–694, 2000.

Lake, J.K., Power, C., and Cole, T.J., Child to adult body mass index in the 1958 British birth cohort: associations with parental obesity, *Arch. Dis. Child.*, 77, 376–381, 1997.

Lake, J.K., *Body Size in Child and Adulthood: Implications for Adult Health* [PhD thesis], University College London, London, 1998.

Li, L., Parsons, T.J., and Power, C., Breast feeding and obesity in childhood: cross sectional study, *BMJ*, 327, 904–905, 2003.

Ong, K.K., Ahmed, M.L., Emmett, P.M., Preece, M.A., and Dunger, D.B., Association between postnatal catch-up growth and obesity in childhood: prospective cohort study [published erratum appears in *BMJ* May 6, 320(7244): 1244], 2000 *BMJ*, 320, 967–971, 2000.

Ong, K.K., Preece, M.A., Emmett, P.M., Ahmed, M.L., and Dunger, D.B., Size at birth and early childhood growth in relation to maternal smoking, parity and infant breast-feeding: longitudinal birth cohort study and analysis, *Pediatr. Res.*, 52, 863–867, 2002.

Parsons, T.J., Power, C., Logan, S., and Summerbell, C.D., Childhood predictors of adult obesity: a systematic review, *Int. J. Obes.*, 23, Suppl. 8, S1–S107, 1999.

Parsons, T.J., Power, C., and Manor, O., Fetal and early life growth and body mass index from birth to early adulthood in 1958 British cohort: longitudinal study, *BMJ*, 323, 1331–1335, 2001.

Parsons, T.J., Power, C., and Manor, O., Infant feeding and obesity through the lifecourse. *Arch. Dis. Child.*, 88, 793–794, 2003.

Power, C. and Moynihan, C., Social class and changes in weight-for-height between childhood and early adulthood, *Int. J. Obes.*, 12, 445–453, 1988.

Power, C. and Matthews, S., Origins of health inequalities in a national population sample, *Lancet*, 350, 1584–1589, 1997.

Power, C., Lake, J.K., and Cole, T.J., Body mass index and height from childhood to adulthood in the 1958 British born cohort, *Am. J. Clin. Nutr.*, 66, 1094–1101, 1997a.

Power, C., Lake, J.K., and Cole, T.J., Measurement and long-term health risks of child and adolescent fatness, *Int. J. Obes.*, 21, 507–526, 1997b.

Power, C. and Jefferis, B.J., Fetal environment and subsequent obesity: a study of maternal smoking, *Int. J. Epidemiol.*, 31, 413–419, 2002.

Power, C., Manor, O., and Matthews, S., Child to adult socioeconomic conditions and obesity in a national cohort, *Int. J. Obes.*, 27, 1081–1086, 2003.

Prentice, A.M. and Jebb, S.A., Obesity in Britain: gluttony or sloth? *BMJ*, 311, 437–439, 1995.

Rasbash, J., Browne, W., and Goldstein, H., *A User's Guide to MLwiN. 2*, Institute of Education, London, 2000.

Rudolf, M.C., Sahota P., Barth, J.H., and Walker, J., Increasing prevalence of obesity in primary school children: cohort study, *BMJ*, 322, 1094–1095, 2001.

Saris, W.H., Blair, S.N., van Baak, M.A., Eaton, S.B., Davies, P.S., Di Pietro, L., Fogelholm, M., Rissanen, A., Schoeller, D., Swinburn, B., Tremblay, A., Westerterp, K.R., and Wyatt, H., How much physical activity is enough to prevent unhealthy weight gain? Outcome of the IASO 1st Stock Conference and consensus statement, *Obes. Rev.,* 4, 101–114, 2003.

Stark, O., Peckham, C.S., and Moynihan, C., Weight and age at menarche, *Arch. Dis. Child.,* 64, 383–387, 1989.

Tanner, J.M., *Growth at Adolescence,* Blackwell Scientific Publications, Oxford, 1962.

Van Lenthe, F.J., Kemper, C.G., and Van Mechelen, W., Rapid maturation in adolescence results in greater obesity in adulthood: the Amsterdam Growth and Health Study, *Am. J. Clin. Nutr.,* 64, 18–24, 1996.

Whichelow, M.J. and Prevost, A.T., Dietary patterns and their associations with demographic, lifestyle and health variables in a random sample of British adults, *Br. J. Nutr.,* 76, 17–30, 1996.

Whitaker, R.C. and Dietz, W.H., Role of the prenatal environment in the development of obesity, *J. Pediatr.,* 132, 768–776, 1998.

Williamson, D.F., Dietary intake and physical activity as "predictors" of weight gain in observational, prospective studies of adults, *Nutr. Rev.,* 54, S101–S109, 1996.

PART III

THE PREVENTION AND TREATMENT OF CHILDHOOD OBESITY

13

CHILDHOOD OBESITY IN THE COMMUNITY: TREATMENT, PREVENTION, AND MONITORING OF CHILDHOOD OBESITY IN LEEDS, U.K.

Mary J. Rudolf

CONTENTS

INTRODUCTION

Childhood obesity has essentially no consequences at the community level and therefore presents a dilemma. On the one hand, communities should be dealing with those health matters that place a real burden on their resources. On the other hand, childhood obesity places almost no immediate burden on resources. However, the burden of adult obesity is highly significant and is predicted by the prevalence of untreated childhood obesity. The medical complications for the vast majority of obese children occur beyond the childhood years. Emotional suffering, which is common, could take up resources, but mental health services are so inadequate across the country that the troubled child is infrequently seen. The community does have a great responsibility to address the problem of childhood obesity, and I propose to address its potential role across the three areas of prevention, treatment, and monitoring.

I use the term "community" advisedly. As professionals, we tend to proffer solutions that are focused within our own spheres of primary care, hospitals, education, or other agencies. Obesity is such a complex and multifaceted problem that I believe we need to think beyond traditional boundaries in order to develop an integrated approach at the community level to tackle this problem.

In this chapter, I shall describe the work that we in Leeds are engaged in, which covers the spectrum of treatment, prevention, and monitoring of childhood obesity. The work is presented, not because we have found ultimate solutions, but more in a spirit of sharing experiences and ideas that can contribute to the national debate.

TREATMENT OF CHILDHOOD OBESITY

In an epidemic of this nature, there is no question that preventive strategies have to be the focus toward which major energies are directed. While acknowledging this, I am struck, as a doctor, by the needs of children who are already obese and the dire lack of services to meet them. The city of Leeds presents a typical picture. There are no specialist pediatric clinics, the hospital dietetic department has closed its doors to any referrals for obesity, and the community dieticians can offer only one or two appointments, with a follow-up phone call. At a meeting of professionals that was called in 2001 because of emerging concerns, it became clear that no additional resources and time were likely to be available, and, dare I say, there was no real inclination from any professional body to take a lead. Clearly, we needed to think "outside the box" for a solution.

The WATCH IT Clinics

The WATCH IT concept grew out of a comment by a colleague that "surely families don't need highly qualified (and expensive) health professionals to help them do what they need to do." The germ of an idea that we might look beyond the traditional medical model was sown, and we warmed to the thought that nonspecialist "clinics" might be the answer, where children and their families could attend and regularly receive basic advice about diet and physical activity, as well as support and encouragement to tackle these.

After great effort, Health Action Zone funding was obtained for a period of 16 months to develop and pilot WATCH IT in disadvantaged areas of South Leeds. Our aim was to develop a community-based program for obese children and their families, which, if promising, would be evaluated by randomized controlled trials.

In January 2003, two community workers were appointed to set up four clinics. The workers were chosen for their personal qualities of empathy and enthusiasm and were instructed to find locations for the service. In November 2003, further funding was obtained from the Leeds Primary Care Research Consortium to establish an additional four clinics in West Leeds run by two further community workers. In response to families' requests, we started to offer group activity sessions in the local leisure centers to supplement the clinics. These sessions offer children and teenagers the opportunity to enjoy physical activity and develop their skills in an environment safe from ridicule.

In WATCH IT, children and their families are seen on an individual basis. The visits focus on the child's well being, and the approach is holistic and flexible to meet individual needs. If emotional and behavioral issues are present (and they usually are), these are addressed. Children are seen as often as weekly, and targets are agreed upon at each session. The key workers receive solid back-up from a professional team, with regular supervision from their team leader and a dietician, psychologist, and sports specialist (see Figure 13.1). Recently, we have linked up with Dr. Deborah Christie, consultant clinical psychologist and senior lecturer at University College London to adapt her HELP approach (Healthy Education Lifestyle Programme) for use by our key workers. This offers families a toolkit of resources to address eating behavior, physical activity, diet, and emotions. It takes a three-pronged approach to develop motivation and externalize worries and concerns and is solution focused (Christie and Viner, in press).

Although at the time of this writing, WATCH IT has only been running for 9 months, both our clinical and research teams are encouraged by progress so far. Fifty-four children, aged 6 to 17 years, have enrolled at the clinics in South Leeds, with the leisure center being the most popular location. Good links have been made with community facilities, a training manual for WATCH IT workers has been written,

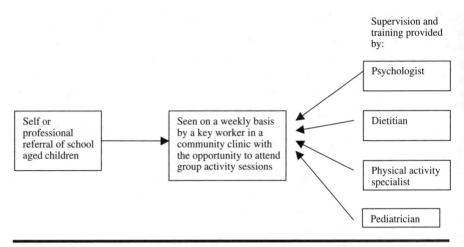

Figure 13.1 The WATCH IT Approach.

and a specialist hospital clinic is being set up to meet the medical needs of the morbidly obese. Attendance has been good and consistent, with remarkably low DNA (did not attend) rates. An independent evaluation in the summer was positive, and parental comments are shown in Box 13.1.

Box 13.1 Parental Comments about the WATCH IT Program

She is more motivated to lose weight

I wish it had been around years ago

The whole family have taken up exercise

I dread to think how she'd be now If she hadn't been coming

Hasn't lost a lot but hasn't put any on. Previously was gaining weight by the week

Consider myself lucky that I heard about it

She has lost weight and feels better about herself   *She has changed her diet completely*

She has really got into her dieting. She can now say NO   *She is happier having seen Jenny and Sheila. She has better body image esteem*

Feels fitter when playing football   *He doesn't get as depressed if he's bullied*

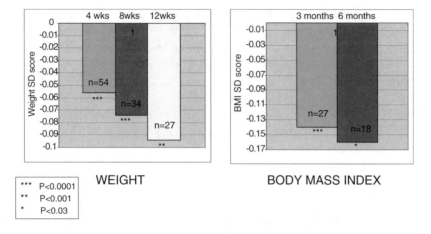

Figure 13.2   Progress to Date of 54 Children Enrolled in the WATCH IT Clinics.

Figure 13.2 shows the progress of the 54 children to October 2003, 27 of whom have been attending for 3 months and 18 for 6 months. There has been a steady, statistically significant decline in weight and body-mass index (BMI). The reduction is not dramatic; however, when considered in the context of the rapid rise in BMI usually seen in children referred to hospital for obesity, it is encouraging, especially because the program is only in the developmental stages.

In light of these results and the interest shown by families, the WATCH IT team is cautiously optimistic that an approach is being developed that is responsive to and addressing the needs of obese children and teenagers in their communities and is maximizing links with local resources. We are clear that an approach like this needs vigorous evaluation by randomized controlled trials to determine its effectiveness; currently, funding is being sought. If WATCH IT is indeed proven to be effective, it offers an exciting model and a low-cost alternative that could be utilized across the country.

PREVENTION OF CHILDHOOD OBESITY

The obesity epidemic is at last recognized as a major public health problem affecting children across the world. As such, it is clear that massive efforts are required to stem the tide. Perhaps, political action will direct the environmental and societal changes that are required in a coherent and effective manner. At present, however, there are a plethora of initiatives being funded across the country. These initiatives are using resources in imaginative ways and are engaging a great deal of creative energy, but little concerted effort is being made to critically evaluate them for their effectiveness. Quite appropriately, there is a demand for action now; rather than wait the years needed to run randomized controlled trials, it is incumbent upon us to seek effective solutions and provide the best quality evidence in a field where little currently exists.

Nine years ago in Leeds, we conducted the first randomized controlled trial in the U.K. of a school-based program for the prevention of obesity risk factors. I describe it now, not because it provided the ultimate answer, but because our experience can be a useful starting point to inform the agenda for action in schools.

The APPLES Program

When APPLES was first conceived, we intended to develop a teaching package to instruct participating schools how to deliver a health-promotion program that would improve pupils' diets and increase their physical activity, to thus reduce their risk of developing obesity (Sahota et al., 2001a). Luckily, we were averted from this medical model by Dr. Rachael Dixey, senior lecturer in health promotion at Leeds Metropolitan University. She convinced us that we should adopt the Health Promoting Schools' philosophy and work with the schools in partnership to help them develop their *own* approach and so have ownership of the project (Sahota et al., 2001b). As researchers, we were naturally concerned that it would be impossible to conduct a trial where each school was doing something different, but we were eventually convinced, and Dr Dixey's approach was undoubtedly responsible for the success of the project.

APPLES was funded by Yorkshire Regional Research and Development. Dr. Pinki Sahota was employed to run the project, engage the schools' participation, provide the input, and evaluate the trial, all with little in the way of additional resources.

Ten schools were recruited and paired for size, ethnicity, and free school meal index. They were then randomized so that five schools received the APPLES program and five were allocated to the control arm. The five control schools received the program 12 months later (see Figure 13.3).

The team met with the schools to help them determine what they wished to tackle. Training days were held, and the delegated teachers were asked to work with colleagues to come up with a school action plan of what they hoped to achieve over the course of the school year. Ideas were offered along with resources, such as dietetic sessions in the classroom, a *Fit is Fun* physical education program, and various teaching materials. The sort of activities that schools chose to use is shown in Box 13.2.

At the end of the year, the project was evaluated in three stages: the implementation process, the impact of the intervention, and finally its effect on the individual. It is essential to know whether the implementation

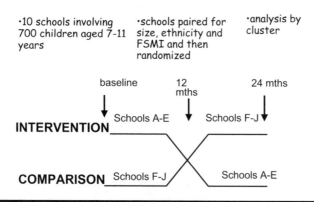

Figure 13.3   Study Design of the APPLES Project, the First Randomized Controlled Trial in the UK of a School-Based Program to Prevent Obesity Risk Factors.

process has been successful when evaluating any complex intervention; if this stage is not successful, it is hardly surprising if the intervention itself is found to be ineffective.

We found that APPLES had been implemented successfully. All 10 schools completed the project. They all developed realistic action plans and achieved most of their action points. There was good uptake of the resources offered, and the teachers' evaluations at the end of the year were positive. Data collection from the children was as high as 65% to 97%, indicating the very high level of commitment of both parents and school staff.

The impact that APPLES had on the schools was quite considerable. Table 13.1 shows the types of action points that the schools opted to take on. By the end of 12 months, 89% of these had been achieved. (Even more gratifying was the finding 5 years later that 65% of them were still

---

Box 13.2   Activities that Schools Chose to Take on as Part of the APPLES Project

- Dietitian visits to class
- PE lessons
- Topic work
- Health Fairs
- Practical cooking sessions
- School Meals
- Snack shops
- Playground activities

Table 13.1 Some Action-Points Undertaken by Schools in Their APPLES Action Plans

	Schools
Nutrition education in the curriculum	10
Healthy eating sessions by dietitian	10
Fit is Fun program in P.E	10
Improved playground facilities	6
Policy changes in break time snacks	5
Healthy snack shops	4

being sustained despite no ongoing input from the APPLES Team.) School meals were a particular case in point, where the quality of meals offered had improved considerably (Table 13.2).

Focus groups were held in all of the schools at the end of 12 months (Dixey et al., 2001a; Dixey et al., 2001b). Facilitators who were blinded to whether the schools were intervention or control led the 80 focus groups of 320 children. They found that APPLES had had a considerable impact on the pupils. They identified that APPLES' pupils had a better understanding about healthy lifestyles and the medical implications. These pupils expressed their understanding in a more sophisticated way and had a higher recollection of activities related to diet and activity. They also reported that their behavior had changed over the year as a result of what they had learned at school.

We were satisfied therefore that APPLES had been implemented and that it had had the desired impact on the schools. The remaining aspect of the evaluation was to ascertain the effects the program had had on the pupils' risk for developing obesity.

Table 13.2 Changes in School Meals Observed Over the Course of 12 Months' Involvement in the APPLES Project

	No. of Schools	
	Before	After
Jacket potatoes	1	10
Fresh fruit—daily	8	10
Mash potatoes	4	6
Salad, vegetables —daily	4	7
Vegetarian options	Poor	Good

The outcome measures we were interested in were:

■ Diet
■ Physical activity
■ Height, weight, and BMI
■ Psychological measures

Our hope was that the children's diet and physical activity levels would improve, as evaluated by 24-hour recall and 3-day diet diaries. BMI and the psychological scales were measured to ensure that we had not induced disordered eating or psychological distress through the program. The data were analyzed by cluster and Figure 13.4 and Figure 13.5 are examples of the data that we obtained.

The results, unfortunately, were disappointing. We found no change in diet and physical activity, other than a very modest (although statistically significant) increase in vegetable intake, as illustrated in Figure 13.4. By the end of 12 months of APPLES, pupils were eating one third of a portion of vegetables more than the control children. Given that baseline vegetable intake was only half a portion, this is an improvement, but hardly the hoped for increase. There were no differences on the psychological measures or in BMI (as shown in Figure 13.5).

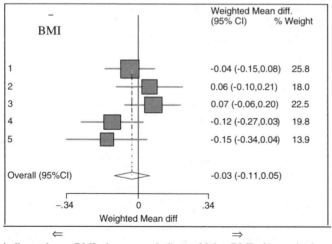

Figure 13.4 Weighted Mean Difference for Vegetable Intake for the Five Intervention Schools Relative to Their Paired Comparison School.

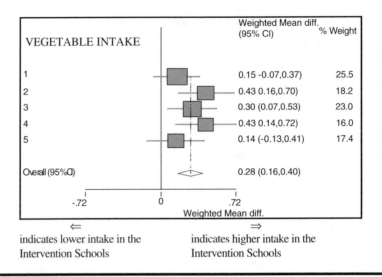

Figure 13.5 Weighted Mean Difference for BMI for the Five Intervention Schools Relative to Their Paired Comparison School.

APPLES therefore did not provide the definitive answer to the prevention of obesity. Perhaps we, and the funding agency, had been overoptimistic about attempting such an ambitious project with few resources. There were good reasons why the trial had not shown APPLES to be effective. The sample size of only 10 schools was too small, the length of intervention of one school year was short, and the available outcome measures were probably not adequate.

However, it was only when we came to analyze the growth data more fully did we realize how naïve we had been. In 1996, there was no awareness about the epidemic of obesity that was taking place. When we analyzed our data over the 3 years that we were in the schools, we uncovered the extent of the problem that we had hoped to tackle. Table 13.3 and Figure 13.6 illustrate the numbers of obese children in the 10 schools between 1996 and 1999. We published these data in the *British Medical Journal* (Rudolf et al., 2001) and were among the first to highlight to the public that we were in the midst of a serious epidemic.

APPLES may not have provided definitive answers in the light of this epidemic, but it did provide some important messages. We found that it was possible, with little in the way of resources, to make a real impact on the school ethos and culture and so foster children's understanding and appreciation of healthy lifestyles. We are now convinced that schools provide an excellent setting for influencing children about diet and activity. However, we are also clear that, in the face of the epidemic's magnitude,

Table 13.3   Percentage of Obese and Overweight Children in 10 Primary Schools in Leeds 1996–1999

Year	N	Mean Age (years)	Overweight[a] (%)	Obese[b] (%)
1996	613	8.39	17.5	8.9
1997	596	9.39	21.3	10.9
1998	577	10.35	25.5	13.9

[a] Defined as BMI >85th centile.

[b] Defined as BMI >95th centile.

far more extensive input and resources are required than are currently available. It is also clear that well-designed research is desperately needed in this area. APPLES may have been the first school-based study in the U.K., but to our knowledge there is no second.

The Leeds Primary Care Research Consortium has recognized the need for more research in this area and has provided us with funding to take APPLES forward. The APPLES trial that we conducted provided some "proof of concept," and we are now actively developing this concept, in particular toward targeting parents more directly, in the hope that we shall receive funding to run a definitive multicenter trial in the future.

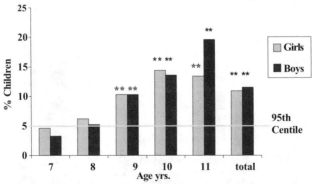

Frequencies significantly different from expected values of 5% (obesity)
* p < 0.01   ** p < 0.001

Figure 13.6   Prevalence of Obesity of Primary School Children According to Age.

MONITORING THE EPIDEMIC
OF CHILDHOOD OBESITY

Having contributed to the public awareness that childhood obesity was a problem, in 2001, we sought to investigate how the APPLES children were doing now that they were in secondary school. Almost 700 children had contributed to the original cohort, and we assumed that the children would now be attending three or so secondary schools in Leeds.

With funding from Regional Research and Development, we attempted to trace the children through primary school leaving lists and the Leeds Education Authority database. The schools were approached, and information sheets were sent to children and parents. In line with our ethics committee guidance, written parental consent was requested to measure the children (Rudolf et al., 2004).

Six hundred and eight children were identified in the leaving lists, of whom only 500 were attending schools close to Leeds. With great effort, we obtained written consent from 348 and eventually measured 338 (68% of those targeted). We were obviously concerned that those children who opted out of being remeasured might be more obese than those who had opted in. However, we were reassured to find, on looking back at our data, that the BMI of our participants was not significantly different from our nonparticipants when they left primary school.

We learned that weight, BMI, and waist circumference measures had continued to rise across the sample in the intervening 3 years since they were last measured (Figure 13.7). Waist circumference measures perhaps sum up the rapidity and extent of the problem.

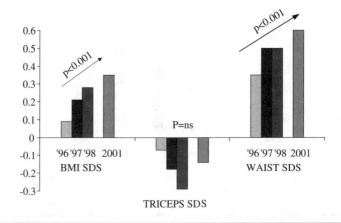

Figure 13.7   Obesity Measures in School Children Between 1996 and 2001.

In 1996, children's waists were on average 1.5 cm larger than expected from the growth references, and this had increased: teenagers were now on average 4 cm larger than expected. This represents an increase of two whole clothes sizes over only 20 years.

The TRENDS Project

These alarming figures have contributed to our awareness that childhood obesity has reached epidemic proportions. As such, it needs to be monitored over time in order to inform us of its progress and to determine whether we are reversing the trend. This is particularly important as multiple efforts are being thrown at the problem, and it is essential that we have a sense as to whether they are having an effect.

Until recently, all children's heights and weights were monitored at school as part of routine child health surveillance; however, during the past several years, this has been discontinued as being wastefulness of resources. The only measures remaining are those taken at school entry, but even here the uptake is poor, and the data are often inaccurate and poorly entered. Universal monitoring of growth and weights, which would allow for screening for obesity, will not return and indeed is not indicated. We have no effective treatment to offer children who are identified as obese, and we could even potentially increase their distress through setting up such a monitoring process and drawing attention to their size. Instead of universal monitoring, what we require is a public health system to ascertain trends across the country and any differences between sociodemographic groups.

We approached the Leeds Primary Care Research Consortium to point out the need for such a study and received a small grant to set up the TRENDS project. Through TRENDS, we are developing a methodology for monitoring the epidemic of obesity in schools and plan to carry out a pilot feasibility study. Thirteen "marker" schools across Leeds have been targeted for this purpose. They have been selected from the five sectors of the city to include five inner-city primary schools with a high proportion of ethnic minority pupils and children entitled to free school meals, five suburban (largely middle class) primary schools, and three high schools.

We propose to measure children at school entry and at ages 9 and 13 years, assessing height, weight, body composition, waist circumference, and BMI, with the intention of returning annually to the schools to measure children at the same ages (rather than a longitudinal cohort). The key to success of this project will be a very high ascertainment of pupils. In view of the need for such a monitoring process and a recognition that figures might well be distorted if obese children fail to volunteer for TRENDS, our ethics committee has taken the unusual step of permitting the project to proceed with opt-out consent rather than with the written parental consent that is usually required.

TRENDS should help us to establish an epidemiological process for monitoring the epidemic of obesity over years and to determine whether political, societal health, and education efforts are having an impact on the problem.

SUMMARY

In this chapter, I have outlined the work that our group is carrying out in Leeds in the areas of treatment, prevention, and monitoring of childhood obesity. We are rigorously evaluating the WATCH IT, APPLES, and TRENDS research projects in the hope that they will contribute to the evidence base of this condition, which to date is rather poor.

We doubt that we will have definitive answers from these projects but hope that, by sharing lessons from this work, we can contribute to building up a body of expertise that will lead to practical and effective interventions that can reverse a process that is damaging the health and well being of increasing numbers of individuals across the developing and developed world.

REFERENCES

Christie, D. and Viner R., HELP! A Healthy Eating and Lifestyle Programme: what, when, where and why overweight, working with adolescents with obesity, *J. Adolesc. Health,* in press.

Dixey, R., Sahota, P., Atwal, S., and Turner, A., Ha ha you're fat, we're strong; a qualitative study of boys' and girls' perceptions of fatness, thinness, social pressures and health using focus groups, *Health Educ.,* 101, 206–216, 2001a.

Dixey, R., Sahota, P., Atwal, S., and Turner, A., Children talking about healthy eating: data from focus groups with three hundred 9–11 year olds, *Br. Nutr. Found. Bull.,* 26, 71–79, 2001b.

Rudolf, M.C.J., Sahota, P., Barth, J.H., and Walker, J., Increasing prevalence of obesity in primary school children: a cohort study, *BMJ,* 322, 1004–1005, 2001.

Rudolf, M.C.J., Greenwood, D.C., Cole, T.J., Levine, R., Sahota, P., Walker, J., Holland, P.C., Cade, J., Truscott, J., Rising obesity and expanding waistlines in school children: a cohort study, *Arch. Dis. Child.,* 89, 235–237, 2003.

Sahota, P., Rudolf, M.C.J., Dixey, R., Hill, A.J., Barth, J.H., and Cade, J., APPLES: a primary school based randomised controlled trial to reduce obesity risk factors, *BMJ,* 323, 1029–1032, 2001a.

Sahota, P., Rudolf, M.C.J., Dixey, R., Hill, A.J., Barth, J.H., Cade, J., APPLES: process and impact evaluation of a primary school based obesity prevention programme in the UK, *BMJ,* 323, 1027–1029, 2001b.

14

OBESITY PREVENTION IN CHILDHOOD AND ADOLESCENCE: A REVIEW OF SYSTEMATIC REVIEWS

John J. Reilly

CONTENTS

INTRODUCTION

Prevalence of pediatric obesity has increased dramatically in recent years across much of the developed world (Reilly and Dorosty, 1999; Strauss and Pollack, 2001). Preschool children (Reilly et al., 1999, Bundred et al., 2001), and children from many parts of the developing world (Ebbeling et al., 2002, Dorosty et al., 2002) have even been affected by the epidemic.

In the U.K. the pediatric obesity epidemic may be gathering pace. In Scotland for example, a recent study (NHS Scotland, 2003) found that prevalence of obesity [defined as body-mass index (BMI) 95th percentile relative to U.K. 1990 population reference data] in 11/12 year olds in 2001 was as high as that of U.S. children of the same age in 1998 (Strauss and Pollack, 2001) using essentially the same definition (Dorosty et al., 2002). Changes in waist circumference in British children and adolescents have been even more marked than changes in obesity prevalence, as defined using BMI percentile (McCarthy et al., 2003). It is also worth noting that the positive energy balance required to produce obesity has probably affected most of the pediatric population: modern children from the U.K. have reduced fat-free mass and greater fat mass compared with their equivalents in the recent past (Wells et al., 2002; Ruxton et al., 1999).

Obesity during childhood and adolescence has important consequences for child and adolescent health and important implications for future (adult) health (Reilly et al., 2003a). High and increasing prevalence of pediatric obesity, combined with comorbidities of obesity, has created a public health crisis (Sokol, 2000; Dietz and Gortmaker, 2001). Addressing this public health crisis requires an emphasis on prevention, and this in turn will need a body of evidence on how best to prevent child and adolescent obesity. Cochrane reviewers (e.g., Campbell et al., 2002) have noted the "mismatch" between the enormous scale of the public health problem and the dearth of high-quality, generalizable evidence of successful intervention strategies for obesity prevention. However, the evidence base in this area is expanding rapidly (Reilly and McDowell, 2003), and a more informed basis for tackling the crisis may be realistic in the near future. The aims of this chapter are as follows:

1. List the recent systematic reviews of interventions for prevention of pediatric obesity.
2. Describe and discuss the principal findings of these reviews.
3. Consider the potential role for modification of physical activity and/or sedentary behavior in prevention.
4. Briefly describe the most promising intervention having encouraging signs of success and based on high-quality evidence.
5. Briefly list ongoing interventions aimed at prevention and consider some lessons for the design and reporting of future interventions.

METHODS

Inclusion/Exclusion Criteria and Search Strategy

The literature search for systematic reviews and meta-analyses described previously (Reilly and McDowell, 2003; Reilly et al., 2002) was updated to October 2003 for the purposes of the present review. The search was cross-checked against two recent systematic reviews, the NHS Centre for Reviews and Dissemination 2002 (NHS CRD, 2002) and a recent Cochrane review (Campbell et al., 2002). The search was also checked against a recent "review of reviews" (Mulvihill and Quigley, 2003).

The aim of the search was to identify systematic reviews and meta-analyses of randomized controlled trials (RCT) in childhood obesity prevention. The accepted hierarchy of evidence, in descending order of importance (reflecting increasing likelihood of bias in conclusions; Harbour and Miller, 2001), is from systematic reviews and meta-analyses of RCT (level 1) to expert opinion (level 4). Expert opinion is particularly prone to bias. Some of the systematic reviews identified by the search set out to synthesize level 1 evidence only. These included the Scottish Intercollegiate Guidelines Network (SIGN) reviews (SIGN 69, 2003; Reilly et al., 2002; Reilly and McDowell 2003) and the NHS Centre for Reviews and Dissemination document 2002 (NHS CRD, 2002).The SIGN review also comprised an evidence-based guideline for the management of childhood obesity.

Other reviews included study designs with lower level evidence such as nonrandomized studies with concurrent control groups, in part because of the dearth of evidence available from RCT. This category of reviews consisted of the NHS Centre for Reviews and Dissemination Report in 1997 (NHS CRD, 1997), which found no RCT; the Cochrane review on obesity prevention (Campbell et al., 2002); and Hardeman et al. (2000). Several other reviews were considered but were less relevant to the present review as their emphasis was on treatment of obesity (usually in clinical samples) rather than prevention. This category included reviews by Epstein et al. (1998) and le Mura and Maziekas (2002).

THE OPTIMAL EVIDENCE BASE

Quality and Quantity of Evidence

To inform strategies for the prevention of childhood and adolescent obesity, a body of consistent, high-quality, generalizable evidence would be ideal. This would require a number of RCT, which were carried out and reported to a high standard. Guidelines for conduct and reporting of RCT are now widely available (Moher et al., 2001) and have been incorporated into guides for the critical appraisal of RCT (e.g., Harbour and Miller, 2001). Describing the characteristics of well-designed and reported RCT is beyond the scope of this chapter but has been described in detail previously in this context (Reilly and McDowell, 2003). Even high-quality RCT may be prone to bias, therefore, a body of RCT suggesting similar conclusions or preferably systematic reviews and/or meta-analyses with a clear conclusion would be ideal. Because the question addressed is obesity prevention, the interventions tested in obesity prevention RCT should have prevention of obesity as a primary aim with at least one objective being a weight-based primary outcome measure.

Follow-Up and Generalizability of Interventions

The issue of duration of study and follow-up is of enormous importance because interventions in this area depend on lifestyle change. A wealth of evidence shows that lifestyle change is often achievable in the short term but cannot be maintained in the longer term. Systematic reviews of meta-analyses that include short-term studies are prone to bias in favor of the intervention for this reason. To minimize this bias, a number of systematic reviews have suggested that follow-up of study participants should last until at least 12 months after the initiation of the intervention, and some reviewers have used this cut-off to exclude or downgrade RCT (e.g., SIGN 69, 2003; Reilly et al., 2002; Reilly and McDowell, 2003; Campbell et al., 2002). An additional issue of great practical importance is the generalizability of particular interventions. This is difficult to assess without replicating interventions in different settings. In the absence of such replication studies, caution is required when considering the applicability of particular interventions to other settings. Most systematic reviews of pediatric obesity prevention and treatment have highlighted this as a major issue for consideration when assessing the evidence, particularly because so much of the published evidence to date has been from the U.S. In childhood obesity treatment RCT for example, the literature is dominated by one single research group from the U.S., thus raising serious questions about generalizability (Reilly et al., 2002).

Other Requirements of the Optimal Evidence Base

Evidence on four additional aspects of any intervention is desirable. First, evaluations of the *process* of the intervention as well as the outcomes are necessary, in particular if we are to understand why and how interventions succeeded or failed. This issue is increasingly being appreciated, and process evaluations are common in more recent obesity prevention RCT (e.g., Sahota et al., 2001a). Second, the cost-effectiveness of the intervention should be considered where possible. This is more difficult to assess, and methodology for assessment of cost-effectiveness is in its infancy. However, cost-effectiveness evaluations are increasingly being regarded as necessary and have begun to be reported in pediatric obesity prevention RCT (Wan-Yang et al., 2003).

Third, intervention studies should ideally measure possible adverse effects of the intervention. The potential for adverse effects of pediatric obesity interventions may have been exaggerated and is probably very limited (Epstein et al., 1998; Sahota et al., 2001b), but possible adverse effects should be considered where possible. Finally, interventions should assess whether changes in behaviors targeted for change have actually occurred, and this requires some attention to the question of how measurable these behaviors are, an issue that is discussed below.

RESULTS

Outcome of Literature Searches

The searches identified nine potentially eligible systematic reviews/meta-analyses (Table 14.1). On closer inspection, one was in fact a "review of reviews" (Mulvihill and Quigley, 2003). Two were focused almost exclusively on treatment rather than prevention (le Mura and Maziekas, 2002; Epstein et al., 1998); three updates of the same reviews of RCT for the SIGN evidence-based guideline on childhood obesity were found (SIGN 69, 2003; Reilly et al., 2002; Reilly and McDowell, 2003); the rest consisted of one NHS CRD review from 1997 (NHS CRD, 1997) and an update published in 2002 (NHS CRD 2002) and one recent Cochrane review (Campbell et al., 2002). It is worth noting that all these systematic reviews have been published in the past 6 years, reflecting the recent increase in awareness of the need for obesity prevention in childhood.

Summary of Each Eligible Review

Table 14.1 summarizes the different reviews and permits a convenient comparison between them. The discussion of the conclusions of the reviews given below is based largely on the later, more recent reviews because these were able to access more evidence. For example, the NHS CRD review published in 1997 (NHS CRD, 1997) has been superseded by the more recent review by the same organization (NHS CRD, 2002). The discussion is also largely based on the reviews of RCT, as the highest level of evidence available.

Main Conclusions of the Reviews

The principal conclusions of each of the eligible reviews are summarized briefly in Table 14.2. All reviews, including the most recent, concluded that there was a remarkable lack of evidence on high-quality, generalizable interventions for prevention of childhood obesity. Simple, effective, and generalizable interventions do not exist at present.

Table 14.1   List of Reviews Identified

Author (Year)	Main Topic	Comments
Systematic Reviews of Obesity Prevention Interventions		
NHS CRD (1997)	Prevention and treatment	Focus on adults; no prevention RCT identified in children
NHS CRD (2002)	Prevention and treatment	Supersedes NHS CRD 1997; review of RCT based on Cochrane review, updated
Campbell et al. (2001)	Prevention	Cochrane review; included RCT and other designs
SIGN 69 (2003) (www.sign.ac.uk) Reilly et al. (2002) Reilly and McDowell (2003)	Diagnosis, prevention, and treatment	Reviews of RCT with critical appraisal; updated to May 2002
Hardeman et al. (2000)	Prevention	Included adults and children; few studies available at that time (RCT or other designs)
Other Reviews		
Epstein et al. (1998)	Treatment	Included RCT and other designs; limited emphasis on prevention
le Mura and Maziekas (2002)	Treatment	Meta-analyses of short-term RCT and other designs; limited emphasis on prevention
Mulvihill and Quigley (2003)	Prevention and treatment	Reviewed some previous systematic reviews

Most of the reviews focused on the potential for targeting reductions in sedentary behavior, particularly TV viewing, in obesity prevention interventions in children and adolescents. Sedentary behavior is at least potentially modifiable (Robinson, 1999), appears to be important to the causation and/or maintenance of obesity, and can be measured objectively with some degree of accuracy and precision (Reilly et al., 2003b). Finally, reduction in TV viewing time is unlikely to do harm and may be beneficial to child health and development in other ways. These issues are important considerations when deciding on which behaviors to target for intervention (Whitaker, 2003; Byers, 2003; Reilly and McDowell, 2003), and sedentary behavior reduction appears to meet these criteria well.

All of the reviews highlighted weaknesses in the design, conduct, and reporting of pediatric obesity prevention RCT. The review by Reilly and

Table 14.2 Principal Conclusions of Systematic Reviews

Author (Year)	Targeting Sedentary Behavior Promising	More Intervention Research Required	No Successful Generalizable Model of Prevention Exists	Need for Improved Methodology	Need for Long-Term Studies	Other Comments
NHS CRD (1997)		✓	✓	—	—	
NHS CRD (2002)	✓	✓	✓	✓	✓	
Campbell et al. (2002)	✓	✓	✓	✓	✓	Summarized RCT in progress
SIGN 69 (2003)	✓	✓	✓	—	✓	Included a management guideline
Reilly et al. (2002)	✓	✓	✓	—	✓	
Reilly and McDowell (2003)	✓	✓	✓	—	✓	Summarized RCT in progress
Hardeman et al. (2000)	—	✓	✓	✓	✓	

McDowell (2003) for example found that only a single RCT was of reasonably high methodological quality (Gortmaker et al., 1999, "Planet Health"). The Cochrane review (Campbell et al., 2002) and the NHS CRD review (NHS CRD, 2002) both concluded that all published RCT had important methodological flaws. It is worth reiterating that these flaws are not merely an academic issue but are likely to bias the conclusions of the studies in question (Moher et al., 2001). One common methodological limitation is short duration of participant follow-up. Reviews have also found that many recent prevention RCT have paid little attention to current guidelines on the design, conduct, and reporting of RCT (Moher et al., 2001). Reviewers have repeatedly concluded that recent RCT have major flaws, including lack of clear aims, inadequate power, inadequate randomization or concealment of random allocation, lack of blinding, differences in treatment of intervention and the control group other than the intervention, high participant attrition and loss to follow-up, no analyses on an intention to treat basis, and high dependence on biased/invalid and imprecise outcome measures.

In summary, although study design and reporting may be improving, the literature on childhood obesity prevention is still dominated by poorer quality, short-term RCT. When the body of highest quality evidence on pediatric obesity prevention is compared against the list of optimal criteria mentioned above, it is clear that there is much work to be done before interventions intended to prevent obesity can be implemented with any confidence. Furthermore, most of the interventions tested in the longer-term individual RCT to date have not actually been successful in reducing obesity risk (e.g., Sahota et al., 2001b; Caballero et al., 2003).

In light of the paucity of high-quality evidence for intervention success, it may be helpful to consider two promising interventions that reported evidence of objectively measured benefit to the intervention and that were well designed. The Planet Health intervention (Gortmaker et al., 1999) was a large-scale, multifactorial, school-based cluster RCT run over two school years in U.S. children of mean age 11.7 years at baseline. The study had a clear aim, was well designed and adequately powered, and used an objective weight-related outcome measure (change in "obesity" prevalence using >85th percentile for BMI and triceps skinfold thickness). A thorough process evaluation was carried out, a very positive although preliminary economic evaluation was recently published (Wan Yang et al., 2003), and possible adverse effects of the intervention were considered. The intervention was intensive, and it targeted decreased sedentary behavior (TV viewing), reduced consumption of energy-dense foods, increases in fruit and vegetable consumption, and increases in "lifestyle" physical activity. This study is also of particular interest because it was unusual in that it reported benefits to the intervention, at least in girls. The authors reported a significant reduction in obesity prevalence in girls in the intervention group (AOR 0.47, 95% CI 0.24 to 0.93) and a significant remission of preexisting obesity in the girls. In boys, reduced obesity prevalence in the intervention group did not reach significance. Gortmaker et al. (1999) reported that the beneficial intervention effects were mediated by reduction in television viewing time. In summary, Planet Health represents a potentially valuable model for school-based obesity prevention programs. It is of particular interest as the only relatively high-quality, long-term intervention to show any benefit of the intervention to date. In addition, process and economic evaluations of the intervention have been encouraging. Further research is currently underway to test the effectiveness of the "Planet Health" approach in other settings in the U.S.

The other RCT that merits a brief summary is that of Robinson (1999). Follow-up in this school-based study was short term (to 7 months after the initiation of the intervention), and this may have biased the outcome in favor of the intervention; however, a longer-term follow-up of the same intervention is underway. The use of a single behavioral target (television viewing) in this study was important. As a result of using a single target for behavioral change, this study demonstrated that this behavior was modifiable and provided important evidence that television viewing/sedentary behavior is *causally* implicated in the development and maintenance of childhood obesity by demonstrating a benefit to the intervention (reduced rate of BMI gain in the intervention vs. control school). As noted above, behavioral targets of interventions should be potentially modifiable and on the pathways that maintain or cause obesity. They should also be measurable with high accuracy and precision and preferably objectively measured. These points may seem obvious but are worth restating.

DISCUSSION

Implications of Present Review

The present "review of reviews" shows that evidence on interventions for the prevention of obesity in children and adolescents is lacking. Even among the higher-quality intervention studies, interventions have usually been unsuccessful. At present, critical appraisal of the evidence leads to the conclusion that we cannot be confident that existing approaches to prevention of child or adolescent obesity have a realistic prospect of success. This conclusion was reached by all of the systematic reviews that have critically appraised the quality of individual studies and that considered their generalizability. Less critical reviews have been more positive about the prospects for obesity prevention, probably because these have been more selective in their literature searching and have placed undue emphasis on a few small, short-term, studies that are biased in favor of the intervention and that in many cases are unlikely to be generalizable.

Ongoing Interventions and Behavioral Targets

Our lack of successful strategies for pediatric obesity prevention has been usefully highlighted by recent systematic reviews and makes the case for more research. A number of obesity prevention RCT are ongoing in children and adolescents (Table 14.3), and these should increase the evidence base markedly in the next few years. Most of these ongoing RCT are based in the U.S., most have focused on targeting modification of sedentary behavior and/or physical activity rather than dietary change, and most have targeted groups perceived to be at high risk of inactivity and/or obesity, such as girls and ethnic minority groups (Kimm et al., 2002). This strategy seems justifiable in view of the evidence suggesting that sedentary behavior is both modifiable and causally related to the development and maintenance of childhood and adolescent obesity (e.g., Gortmaker et al., 1999; Robinson, 1999). A meta-analysis of childhood obesity *treatment* interventions, which compared diet vs. diet plus "exercise" (this term included lifestyle physical activity, structured aerobic exercise, and modification of sedentary behavior) concluded that effects of exercise were additive to those of diet (Epstein and Goldfield, 1999). This review also concluded that addition of "exercise" to protocols for treatment of childhood overweight and obesity was particularly beneficial in modifying comorbidities of obesity, such as cardiovascular risk factors (Epstein and Goldfield, 1999). In childhood obesity prevention, most of the older studies have attempted to modify both diet and physical activity and separating effects of each type of intervention has not been possible to date. However, as noted above, interventions that exclusively target television viewing and target no other behavioral changes have been promising; even in the more complex multifaceted interventions, it appears that modification of television viewing has mediated the success of the intervention (Robinson, 1999; Gortmaker et al., 1999).

With the high prevalence of overweight and obesity, a surprisingly high prevalence of cardiovascular risk factors among overweight and obese children (Freedman et al., 1999; Reilly et al., 2003a), and the fact that high degrees of sedentary behavior are a major cardiovascular risk independent of obesity, the potential for public health gain from modifying sedentary behavior across the population is immense. This is particularly important because, in contemporary populations, sedentary behavior seems to be established at an early age (Reilly et al., 2004; Certain and Kahn 2002).

In general, studies that have directly targeted dietary changes in children have shown that these behaviors are less modifiable than sedentary behaviors. Dietary interventions are popular but carry the risk of wasted intervention effort in targeting behaviors that might be the most resistant to change. Some dietary behaviors that have been targeted in past interventions are not even clearly related to obesity development or maintenance (Reilly and McDowell, 2003). It should also be noted that the

Table 14.3 Ongoing Obesity Prevention RCT

Principal Investigator	Trial Name/Acronym	Start Date/End Date	Participants	Notes/Further Details
	NHLBI 2001, Decreasing adiposity in African-American pre-adolescent girls			www.nhlbi.nih.gov/resources/docs/plandisp.htm
Dr T. Robinson	Long-term follow-up of reducing children's TV viewing to prevent obesity	1999 to 2002	3rd grade pupils in public elementary schools, U.S.	Tom.robinson@stanford.edu
Dr. June Stevens	Trial of activity for adolescent girls (TAAG)	2001 to 2007	Adolescent girls, U.S.	June-Stevens@unc.edu
Dr. J. Reilly	Movement and Activity Glasgow Intervention in Children (MAGIC)	2002 to 2005	Children in their preschool year, Scotland	jlr2y@clinmed.gla.ac.uk
Dr. T. Robinson	Girls health enrichment multisite studies (GEMS)		African-American preadolescent girls, U.S.	Tom.robinson@stanford.edu
Dr. T. Robinson	Stanford Games		7- to 10-year-old girls in after-school care, U.S.	Tom.robinson@stanford.edu

Sources: From Campbell et al., *The Cochrane Library*, Update Software, Oxford, issue 2, 2002, and Reilly and McDowell, *Proc. Nutr. Soc.*, 62, 611–619, 2003; Material supplied by Dr. Selvi Williams, author of ongoing review of prevention RCT.

mechanisms of success in interventions that target reductions in television viewing are not clear and may be complex. Reducing time spent in sedentary behavior for obesity treatment and prevention may be successful not because this increases physical activity but because it might reduce food intake. Children and adolescents who reduce their television viewing time may experience a less positive energy balance than controls either because this alters their energy intake or because this increases their energy expenditure, or perhaps a combination of both (Reilly and McDowell, 2003).

Hardeman et al. (2000) observed that many interventions identified in their systematic review had no theoretical basis and lacked an emphasis on the methods of encouraging behavioral change that are effective in other contexts. Future interventions should at least consider using methods of behavior change that have been promising in other settings. Simply providing information/health education materials is usually insufficient to bring about sustained behavioral changes to prevent (Hardeman et al., 2000) or treat (Reilly, 2002) chronic diseases of energy imbalance.

Methodological Issues in Intervention Studies

Future intervention studies should also consider the impact of limitations in methodology on study design and interpretation. Recent systematic reviews have consistently concluded that imprecise and inaccurate measures of exposure and outcome variables such as dietary intake and physical activity have hampered the ability of investigators to test the efficacy of their interventions. (Hardeman et al., 2000; Campbell et al., 2002; Summerbell et al., 2003; Reilly and McDowell, 2003). More objective methods can now offer higher precision and accuracy of some variables, particularly physical activity and sedentary behavior (Reilly et al., 2003b; Hardeman et al., 2000). They can also reduce the self-reporting biases that tend to afflict dietary interventions (Byers, 2003). Most dietary variables are probably not measurable with sufficient accuracy or precision, and biases arising from self reporting of dietary intake have probably been common. This issue was discussed in more detail in a recent editorial (Byers, 2003) on the "Pathways" obesity prevention trial (Caballero et al., 2003). Byers (2003) suggested that bias in self reporting of dietary intake and physical activity was probably responsible for the apparent benefit of the intervention in these behaviors in the absence of any detectable effect on the more objective outcome measures of body weight or composition.

Using more objective measures of physical activity in studies on the scale of Pathways ($n = 1704$ children in 41 schools) would be challenging, but this may be essential in future studies of this kind. Placing less emphasis on poorly measured (dietary) variables should also free research time and study resources for collecting outcome data on objectively measured variables and may reduce the burden of measurement on study participants. Dietary assessment is time consuming for subjects and tends to be associated with poor data collection rates. The burden of dietary assessment may encourage participant dropout. The school-based "APPLES" intervention in Leeds, U.K. (Sahota et al., 2001b) found that one dietary assessment method showed a small but statistically significant increase in fruit and vegetable consumption in the intervention schools, but this was not observed with a second dietary assessment method and was probably too small to be of biological significance. Subject adherence to the dietary assessment was poor, and dropout rates for the dietary assessment outcomes were high (Sahota et al., 2001b). In view of these methodological problems and the apparently limited potential to modify dietary intake using the school- and family-based approaches, the importance of including dietary intake (modification and measurement) in obesity prevention studies must be questioned.

When to Intervene for Obesity Prevention?

The preceding section dealt with the topic of *how* to intervene to prevent obesity in children and adolescents. Most observers have neglected the issue of *when* to intervene. It is first important to note that the pathways that lead to obesity in adults can have early-life origins (Whitaker, 2002). A number of *critical periods* during which later risk of obesity might be "programmed" have now been described: life in utero, infancy, the preschool period before the adiposity rebound, and adolescence (Dietz, 1997). The mechanisms of programming during these periods are generally poorly understood at present (Whitaker and Dietz, 1998), and a systematic review found many weaknesses in observational studies that attempted to test for associations between early-life events and adult obesity (Parsons et al., 1999). Nevertheless, the body of evidence suggesting that these periods are critical is increasing rapidly in both quantity and quality, and the evidence on mechanisms is also increasing. More recent studies have identified several early-life risk factors as possible candidates for intervention. These include factors that might influence later obesity risk in utero, such as maternal smoking during pregnancy (Von Kries et al., 2002). Factors that operate in infancy may also be important. Breast-feeding may be protective against later obesity risk (Dietz, 2001; Armstrong et al., 2002). Evidence is not consistent, although larger and more recent studies with better measures of breast-feeding exposure tend to show a degree of consistency. Rate of weight gain during infancy also appears to be an important independent risk factor for later obesity (Stettler et al., 2002).

The "adiposity rebound" is also of particular interest; earlier adiposity rebound, when BMI begins to increase after its nadir in early childhood (Rolland-Cachera et al., 1984), confers much greater risk of adult obesity

(Whitaker et al., 1998). There is a marked recent secular trend to earlier adiposity rebound in Europe (Reilly et al., 2001) that may have contributed to the adult obesity epidemic. Evidence on factors that influence timing of the adiposity rebound is scarce at present, but dietary influences seem unlikely (Dorosty et al., 2000), implying that the energy expenditure side of the energy balance equation may be more influential (Dorosty et al., 2000). From a public health perspective, shifting the timing of adiposity rebound back toward that characteristic of the recent past (perhaps by increasing physical activity and/or decreasing sedentary behavior in young children) might conceivably have great benefits by "reprogramming" the regulation of energy balance and body composition across a large segment of the population. This might be more feasible than attempting to modify physical activity and sedentary behavior later in life, as young children appear to have greater behavioral flexibility and may be better disposed toward physical activity (Reilly and McDowell, 2003). Even if the period of the adiposity rebound is not strictly "critical" in terms of nutritional programming, modifying physical activity and sedentary behavior in early childhood might help reduce later obesity risk by establishing a less "obesogenic" lifestyle at an early stage in life when later lifestyle habits may be forming.

In the near future, it should be possible to consider interventions for the prevention of later obesity that are focused on particular events or behaviors in early life. These would have a number of advantages. First, the critical "window" for intervention for the programs would be relatively short. For example, for maternal smoking during pregnancy or promotion of breast-feeding, intervention effort could be concentrated on a few months of life in each individual. Second, such interventions could bring about other benefits: reduced smoking during pregnancy would benefit mother and baby in other respects (as would breast-feeding). At the very least, these interventions should do no harm if carried out sensitively and should have the potential to reach much of the population in early life — population-based strategies are essential if the obesity epidemic is to be tackled successfully.

The systematic reviews summarized for this chapter found no evidence of early-life interventions that aimed to prevent later obesity. These periods have therefore been neglected in the obesity prevention literature but clearly offer some promise for interventions across the population if longer-term research funding (to follow participants from early life to adulthood) can be obtained, particularly in view of the apparent failure of interventions later in life. Carrying out such interventions would require a major change in philosophy among funding bodies and researchers, with much greater recognition of the importance of the early life environment to later nutritional health and disease. Designing studies that evaluate such interventions would not be straightforward, and the "gold standard" RCT may not be practical or ethical, but alternative designs are available.

CONCLUSIONS

The present review of systematic reviews demonstrates that the evidence base on childhood obesity prevention has increased markedly in recent years. However, the evidence base remains extremely limited, and no successful, high-quality, generalizable interventions presently exist. The existing systematic reviews provide useful directions as to the nature, design, and reporting of future RCT in obesity prevention, which are required so urgently. The ongoing RCT in childhood obesity prevention listed in this chapter will produce valuable evidence in 5 to 10 years. In the meantime, interventions should probably be based on the principles outlined by Whitaker (2003). Whitaker argued that interventions should meet four criteria: (1) they should do no harm, (2) they should bring about improvements in child health and development independent of obesity, (3) they should target behaviors that have evidence showing that they are likely related to the development and/or maintenance of obesity, and (4) they should target behaviors that are modifiable. This cautious approach provides a useful framework for the design of future interventions, and these four tests should be applied when such interventions are being planned. Whitaker (2003) also argued that there was reasonable evidence that these four tests were met by modifications targeting reduced sedentary behavior, promotion of breast-feeding, and reduction of sugar-sweetened drink consumption. From the point of view of designing trials with outcomes that can be measured successfully, in order that future interventions can be evaluated, one additional criterion may be helpful: interventions should target behaviors that are measurable with limited scope for imprecision and self-reporting bias. In practice, this will probably mean a greater dependence in future intervention studies on *objective* measures of intervention outcome (Hardeman et al., 2000; Reilly and McDowell, 2003).

REFERENCES

Armstrong, J. and Reilly, J.J., Child Health Information Team, Breastfeeding and lowering the risk of childhood obesity, *Lancet,* 359, 2003–2004, 2002.

Bundred, P., Kitchiner, D., and Buchan, I., Prevalence of overweight and obese children between 1989–1998: population based series of cross-sectional studies, *BMJ,* 322, 1–4, 2001.

Byers, T., On the hazards of seeing the world through intervention-colored glasses, *Am. J. Clin. Nutr.,* 78, 904–905, 2003

Caballero, B., Clay, T., Davis, S.M., Ethelbah, B., Rock, B.H., Lohman, T.G., Norman, J., Story, M., Stone, E.J., Stephenson, L., and Stevens, J., Pathways: a school-based, randomised controlled trial for the prevention of obesity in American Indian school children, *Am. J. Clin. Nutr.,* 78, 1030–1038., 2003

Campbell, K., Waters, E., O'Meara, S., Edmunds, L., Kelly, S., and Summerbell, C., Interventions for preventing obesity in children (Cochrane review), in *The Cochrane Library,* Update Software, Oxford, issue 2, 2002.

Certain, L.K. and Kahn, R.S., Prevalence, correlates, and trajectory of TV viewing among infants and toddlers, *Pediatrics*, 109, 634–642, 2002.

Dietz, W.H., Periods of risk for the development of childhood obesity: what do we need to learn? *J. Nutri.*, 127, s1884–1886, 1997.

Dietz, W.H., Breastfeeding may help prevent childhood obesity, *JAMA*, 285, 2501–2507, 2001.

Dietz, W.H. and Gortmaker, S.J., Preventing obesity in children and adolescents, *Annu. Rev. Public Health*, 22, 337–353, 2001.

Dorosty, A.R., Rogers, I.S., Emmett, P.M., and Reilly, J.J., Factors associated with early adiposity rebound, *Pediatrics*, 105, 1115–1118, 2000.

Dorosty, A.R., Siassi, F., and Reilly, J.J., Obesity in Iranian children, *Arch. Dis. Child.*, 87, 388–391, 2002.

Ebbeling, C.B., Pawlak, D.B., and Ludwig, D.S., Childhood obesity: public health crisis, common sense cure, *Lancet* 360, 473–482, 2002.

Epstein, L., Myers, M.D., Raynor, H.A., and Saelens, B.E., Treatment of paediatric obesity. *Pediatrics*, 101, 554–570, 1998.

Epstein, L.H. and Goldfield, G.S., Physical activity in the treatment of childhood overweight and obesity: current evidence and research issues, *Med. Sci. Sports Exerc.*, 31, (Suppl. 1), s553–559, 1999.

Freedman, D.S., Dietz, W.H., and Srinivasan, S.R., The relation of overweight to cardiovascular risk factors among children and adolescents: the Bogalusa Heart Study, *Pediatrics*, 103, 1175–1182, 1999.

Gortmaker, S.L., Peterson, K., Wiecha, J., Sobol, A.M., Dixit, S., Fox, M., and Laird, N., Reducing obesity via a school-based interdisciplinary intervention among youth, *Arch. Pediatr. Adolesc. Med.*, 153, 409–418, 1999.

Harbour, R. and Miller, J., A new system for grading recommendations in evidence based guidelines, *BMJ*, 323, 334–336, 2001.

Hardeman, W., Griffin, S., Johnston, M., Kinmonth, A.L., and Wareham, N.J., Interventions to prevent weight gain: a systematic review of psychological models and behaviour change methods, *Int. J. Obes.*, 24, 131–143, 2000.

Kimm, S.Y.S., Glynn, N.W., Kriska, A.M., Barton, B.A., Kronsbert, S.S., Daniels, S.R., Crawford, P.B., Sabry, Z.I., and Kiang, L., Decline in physical activity in black girls and white girls during adolescence, *N. Engl. J. Med.*, 347, 709–715, 2002.

le Mura, L.M., and Maziekas, M.T., Factors that alter body fat, body mass, and fat-free mass in pediatric obesity, *Med. Sci. Sports Exerc.*, 34, 487–496, 2002.

McCarthy, D., Ellis, S.M., and Cole, T.J., Central overweight and obesity in British youth aged 11–16 years, *BMJ*, 326, 624–629, 2003.

Moher, D, Schulz, K.F., and Altman, D.G., The CONSORT statement, *Lancet*, 357, 1191–1194, 2001.

Mulvihill, C. and Quigley, R., *The Management of Obesity and Overweight: An Analysis of Reviews of Diet, Physical Activity and Behavioural Approaches*, Evidence briefing, 1st ed., Health Development Agency, London, 2003.

NHS CRD, Centre for Reviews and Dissemination, *A Systematic Review of the Interventions for the Prevention and Treatment of Obesity, and the Maintenance of Weight Loss*, CRD Report 10, University of York, 1997.

NHS CRD, The prevention and treatment of childhood obesity, In *Effective Health Care* 7 (6), 2002.

NHS Scotland, *Quality Improvement Indicators 2003*, 2003.

Parsons, T.J., Power, C., Logan, S., and Summerbell, C.D., Childhood predictors of adult obesity: a systematic review, *Int. J. Obes.* 23 (Suppl. 8), s1–107, 1999.

Reilly, J.J. and Dorosty A.R., Epidemic of obesity in UK children, *Lancet,* 354, 1874–1875, 1999.

Reilly, J.J., Dorosty, A.R., and Emmett, P.M., Prevalence of overweight and obesity in British children: cohort study, *BMJ,* 319, 1039, 1999.

Reilly, J.J., Kelly A., Ness P., Dorosty A.R., Wallace W.H.B., Gibson B.E.S., and Emmett, P.M., Premature adiposity rebound in children treated for acute lymphoblastic leukaemia, *J. Clin. Endocrinol. Metab.,* 86, 2775–2778, 2001.

Reilly, J.J., Understanding chronic malnutrition in childhood and old age: role of energy balance research, *Proc. Nutr. Soc.,* 61, 321–327, 2002.

Reilly, J.J., Wilson, M.L., Summerbell, C.D., and Wilson, D.C., Obesity diagnosis, prevention and treatment; evidence-based answers to common questions, *Arch. Dis. Child.,* 86, 392–394, 2002.

Reilly, J.J. and McDowell, Z.C., Physical activity interventions in the prevention and treatment of paediatric obesity: systematic review and critical appraisal, *Proc. Nutr. Soc.,* 62, 611–619, 2003.

Reilly, J.J., Methven E., McDowell S.C., Hacking B., Alexander P., Stewart L., and Kelnar, C.J.H., Health consequences of obesity: systematic review, *Arch. Dis. Child.,* 88, 748–752, 2003a.

Reilly, J.J., Coyle, J., Kelly, L.A., Burke, G.B., Grant, S., and Paton, J.Y., An objective method for measurement of sedentary behaviour in 3–4 year olds, *Obes. Res.,* 11, 1155–1158, 2003b.

Reilly, J.J., Jackson, D.M., Montgomery, C., Kelly, L.A., Slater, C., Grant, S., and Paton, J.Y., Total energy expenditure and physical activity in young Scottish children: mixed longitudinal study, *Lancet,* 363, 211–212, 2004.

Robinson, T.N., Reducing children's television viewing to prevent obesity: a randomised controlled trial, *JAMA,* 282, 1561–1567, 1999.

Rolland-Cachera, M.F., Sempe, M., and Guillard-Battaille, M., Adiposity rebound in children, *Am. J. Clin. Nutr.,* 39, 129–135, 1984.

Ruxton, C.H.S., Reilly, J.J., and Kirk, T., Body composition of healthy 7–8 year-olds and a comparison with the "reference child," *Int. J. Obes.,* 23, 1276–1279, 1999.

Sahota, P., Rudolf, M.C.J., Dixey, R., Hill, A.J., Barth, J.H., and Cade, J., Evaluation of implementation of a primary school based intervention to reduce risk factors for obesity, *BMJ,* 323, 1027–1029, 2001a.

Sahota, P., Rudolf, M.C.J., Dixey, R., Hill, A.J., Barth, J.H., and Cade, J., Randomised controlled trial of primary school-based intervention to reduce risk factors for obesity, *BMJ,* 323, 1–5, 2001b.

Scottish Intercollegiate Guidelines Network (SIGN), *SIGN 69, Management of Obesity in Children and Young People: A National Clinical Guideline,* available at www.sign.ac.uk, 2003.

Sokol, R.J., The chronic disease of childhood obesity: the sleeping giant has awakened, *J. Pediatr.,* 136, 711–713, 2000.

Stettler, N., Zemel, B.S., Kumanyika, S., and Stallings, V.A., Infant weight gain and childhood overweight status, *Pediatrics,* 109, 194–199, 2002.

Strauss, R.S. and Pollack, H.A., Epidemic increase in childhood overweight, *JAMA,* 286, 2845–2848, 2001.

Summerbell, C.D., Waters, E., Edmunds, L., and Kelly, S., Interventions for treating obesity in children, in *The Cochrane Library,* Update Software, Oxford, issue 2, 2003.

Von Kries, R., Toschke, A.M., Koletzsko, B., and Slikker, W., Maternal smoking during pregnancy and childhood obesity, *Am. J. Epidemiol.,* 156, 954–961, 2002.

Wan-Yang, L., Yang, Q., Lowry, R., and Wechsler, H., Economic analysis of a school-based obesity prevention program, *Obes. Res.,* 11, 1313–1324, 2003.

Wells, J.C.K., Coward, W.A., Cole, T.J., and Davies, P.S.W., The contribution of fat and fat-free tissue to body mass index in contemporary children and the reference child, *Int. J. Obes.* 26, 1323–1328, 2002.

Whitaker, R.C. and Dietz, W.H., The role of the pre-natal environment in the development of obesity, *J. Pediatr.,* 132, 768–776, 1998.

Whitaker, R.C., Pepe, M.S., Wright, J.A., and Dietz, W.H., Early adiposity rebound and risk of adult obesity, *Pediatrics,* 101, E5, 1998.

Whitaker, R.C., Understanding the complex journey to obesity in early adulthood, *Arch. Int. Med.,* 136, 923–925, 2002.

Whitaker, R.C., Obesity prevention in primary care: four behaviours to target, *Arch. Pediatr. Adolesc. Med.,* 151, 725–727, 2003.

15

THE BARRIERS TO, AND THE
FACILITATORS OF, HEALTHY
EATING AMONG CHILDREN:
FINDINGS FROM A
SYSTEMATIC REVIEW

*J. Thomas, K. Sutcliffe, A. Harden, A. Oakley,
S. Oliver, R. Rees, G. Brunton, and J. Kavanagh*

CONTENTS

INTRODUCTION

Encouraging children to eat healthily is advocated in the belief that they will benefit from the long-term physiological consequences of a good diet and that establishing healthy eating patterns early in life will lead to life-long healthy eating. Fruits and vegetables have been singled out as an important component of a healthy diet in all age groups (Department of Health, 2000; World Health Organization, 2003).

Recent surveys have found that British children are eating less than half the recommended number of portions of fruits and vegetables per day (Department of Health, 2000; Doyle and Hosfield, 2003), and health promotion interventions involving education and/or environmental changes have been identified as an important route toward increasing children's intake of fruits and vegetables. This chapter describes a systematic review that addressed the question of whether such interventions increase fruit and vegetable intake, along with questions concerning children's views and experiences of healthy eating.

Systematic reviews have the explicit aim of avoiding drawing wrong or misleading conclusions either from biases in the review or from biases in the studies contained in the review. Reviewers achieve this by searching extensively for relevant research, assessing the validity and reliability of this research, and basing their conclusions only on the findings of the research in the review (Cooper and Hedges, 1994). Systematic reviews have developed in response to a growing need for policymakers and practitioners to have "short cuts" to the latest research evidence when making decisions. The question for our review was decided by a steering group, which included relevant policymakers at the Department of Health (U.K.), other specialists in health promotion, and experts in research synthesis. The rationale for undertaking the review, therefore, was a practical one: to inform future decisions in this area by policymakers and practitioners.

A number of previous systematic reviews target healthy eating, although none have focused specifically on increasing fruit and vegetable consumption among children (Centre for Reviews and Dissemination, 2002; Ciliska et al., 2000; Contento et al., 1992; Fulton et al., 2001; Hursti and Sjödén, 1997; Lytle, 1994; McArthur, 1998; White et al., 1998). Our review aimed to address this gap but also aimed to provide a richer context for interpreting the success or otherwise of interventions by acknowledging that children's own understandings and experiences of health and social issues are a valuable, but neglected, resource for intervention development (Brannen et al., 1994; Moore and Kindness, 1998; Peersman, 1996; Shucksmith and Hendry, 1998). Therefore, in addition to randomized controlled trials (RCTs) and nonrandomized controlled trials (trials), we also included "qualitative" findings from studies assessing children's perspectives on healthy eating ("views" studies). This chapter describes the methods of the review briefly before concentrating on its results.*

METHODS

Figure 15.1 outlines our approach to the review. We began with a broad question — related to healthy eating in general — and searched extensively using established methods for systematic reviews in health promotion. These included electronic searches, looking through journals manually ("hand searching"), and contacting authors and relevant organizations (both in the U.K. and internationally). Searches were very inclusive at this stage in order to identify studies evaluating the effects of interventions as well as those examining children's views. Both published and unpublished studies were sought. We used the results of this search to draw up a "map" of research activity showing where research had been conducted, what it examined, and the type of methodology it employed. The map was used to inform a consultation process with our key stakeholders, which resulted in the focus for our in-depth review: fruit and vegetable consumption. We consider that obtaining the views of the people who will ultimately be using our reviews to be very important. If systematic reviews are to assist policymakers and practitioners to make decisions effectively, they must not only address current concerns of these groups of people but do so in a relevant and applicable way.

Both trials and "views" studies had to include children whose average age was between 4 and 10 years old (the specific age focus was selected by our advisory group). In addition,

1. Trials (both randomized and nonrandomized) were included if they:
   ■ Were reported in the English language;

---

* A full technical report (Thomas et al., 2003) is available online at: http://eppi.ioe.ac.uk/.

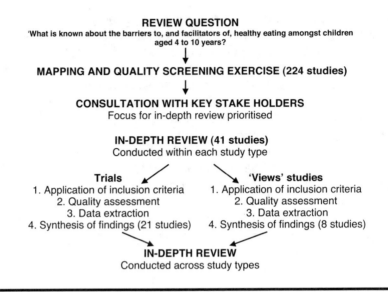

REVIEW QUESTION
'What is known about the barriers to, and facilitators of, healthy eating amongst children aged 4 to 10 years?

MAPPING AND QUALITY SCREENING EXERCISE (224 studies)

CONSULTATION WITH KEY STAKE HOLDERS
Focus for in-depth review prioritised

IN-DEPTH REVIEW (41 studies)
Conducted within each study type

Trials
1. Application of inclusion criteria
2. Quality assessment
3. Data extraction
4. Synthesis of findings (21 studies)

'Views' studies
1. Application of inclusion criteria
2. Quality assessment
3. Data extraction
4. Synthesis of findings (8 studies)

IN-DEPTH REVIEW
Conducted across study types

Figure 15.1   Overall Approach Taken in the Review.

- Measured fruit and/or vegetable intake or fruit and/or vegetable related knowledge, attitudes, or intentions; and
- Employed a control or comparison group.
2. "Views" studies were included if they:
   - Were conducted in the U.K.;
   - Described children's views about fruit and/or vegetables;
   - Were reported in or after 1990; and
   - Reported at least some information on all of the following: the research question, procedures for collecting data, sampling and recruitment methods, and at least two sample characteristics.

The internal validity of both sets of studies was assessed using tools designed for each specific type of study. Two reviewers worked independently on this and then met to compare their results; they were not blinded to study findings of authors during this process.

There are few established methods for bringing together (or *synthesizing*) the results from diverse types of studies such as those identified for this review. In recognition of this, new methods were developed for synthesizing the "views" studies alone and for bringing them together with the trials. The review contained three syntheses: a statistical meta-analysis to combine the trials, a thematic analysis to combine the "views" studies, and a cross-study synthesis to bring both types of studies together. The methods are described in more detail elsewhere (Thomas et al., 2004), and the challenges and implications of combining qualitative and quantitative research are discussed in a further paper (Harden and Thomas, in press).

RESULTS: "QUANTITATIVE" SYNTHESIS OF TRIALS (STATISTICAL META-ANALYSIS)*

After we assessed the studies for their methodological quality, 11 of 33 outcome evaluations were excluded from the synthesis. Five of these studies were excluded due to methodological weaknesses as judged by our quality assessment tool (Peersman et al., 1997). Six other studies were excluded on our methodological filter: either they only allocated one cluster of children to intervention and comparison groups** ($n = 5$), or it was not possible to calculate an effect size from the data presented ($n = 1$). Three of the remaining studies focused on measures designed to persuade children to try new vegetables, rather than increasing their overall intake; their results are described later. Figure 15.2 shows the results of this filtering process.

Nineteen studies were included in our meta-analysis. These were further subdivided into those studies that were rated as having high methodological quality and those that were rated medium. The difference between "high"- and "medium"-quality studies was usually due to the way in which their results were reported. For example, the equivalence of intervention and control groups is an important part of the internal validity of a trial. Studies that presented data to demonstrate either that their groups were equivalent or that they had taken account of differences in their analysis would have been rated higher than those that merely stated that there were no differences between groups. The latter studies were marked for testing in a sensitivity analysis before their findings could contribute to the synthesis. However, no significant differences were detected between the two groups of studies (or between individual studies and the rest of the group), although there was a suggestion that the nonrandomized studies showed higher effects on some outcomes than those that had employed random allocation.

Statistical methods were employed to pool the results of the trials. Studies addressing the same outcomes were identified (knowledge, attitudes, behavioral measures), and, if statistical tests revealed no significant heterogeneity,

---

* The statistical results of the meta-analysis are available online (http://eppi.ioe.ac.uk), and this link also includes a detailed breakdown of the quality assessment process.

** This issue is important because, for example, children attending the same school are likely to have similar characteristics. This affects the validity of statistical tests. For further reading on the allocation of clusters of individuals in research studies, please see Murray (1998).

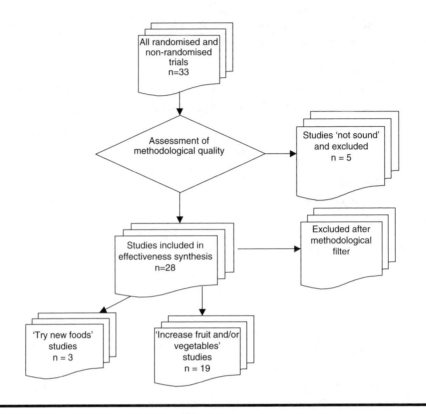

Figure 15.2 Flow of Studies Through the Quantitative Synthesis.

their data were pooled, and an overall effect size was calculated. In addition to pooling effect sizes, the differences between studies were compared on a standardized scale in order to evaluate the relative effectiveness of each intervention.

The types of interventions evaluated by these studies were largely school-based and often combined learning about the health benefits of fruit and vegetables with "hands-on" experience in the form of food preparation and taste-testing. The majority targeted parents and/or involved them in intervention delivery alongside teachers and health promotion practitioners. Some included environmental modification involving, for example, changes to the foods provided at school. Some interventions targeted more than one outcome (for example, fruit and vegetable consumption, fat intake, knowledge, self-efficacy, body-mass index (BMI), and physical activity).

Because all but two of the studies contained different interventions, we placed more emphasis on a narrative describing the studies rather than the summary statistic obtained from the meta-analysis.

Can Interventions Increase Fruit and Vegetable Consumption?

Pooled estimates from the 19 studies suggested that implementation of these interventions will, on average, increase children's fruit intake by one fifth of a portion per day and their vegetable intake by a little less then one fifth of a portion per day. These are averages however, and different interventions will produce different effects. Bigger effects are associated with targeted interventions for parents with risk factors for cardiovascular disease (increasing fruit and vegetable intake by almost two portions) and with those interventions that do not "dilute" their focus on fruit and vegetables by trying to promote physical activity or other forms of healthy eating (for example, reduced intake of sodium and fat) in the same intervention (pooled effect sizes were three times higher in these studies). Single component interventions, such as classroom lessons alone or providing fruit-only snacks, were not effective.

Two main messages emerged from the findings of 14 studies that conducted integral process evaluations: (1) promoting healthy eating can be an integral and acceptable component of the school curriculum, and (2) effective implementation in schools requires skills, time, and support from a wide range of people.

From the Health Survey for England 2001 (Doyle and Hosfield, 2003) undertaken on behalf of the Department of Health, we know that children consume, on average, between 2.5 and 2.8 portions of fruit and vegetables per day. Only one study in the present review (Epstein et al., 2001) was able to increase fruit and vegetable consumption by enough to approach the recommended daily consumption of five portions. Only one of the other interventions was able to increase daily intake by as much as one portion (Auld et al., 1999). In our review, the median effect size* for fruit and vegetable consumption was 0.35 with a mean of 0.32. The effect size necessary to increase consumption from 2.5 to 5 is approximately equal to 1.2. In their study of 302 psychological, educational and behavioral interventions, Lipsey and Wilson (2001) found that the median effect size was 0.47 and the mean was equal to 0.5. This suggests that the studies included in our review were doing, in general, not quite as well as those in other fields.

Thus, judging from the experiences of a large number of other studies, only an exceptionally successful intervention would be able to increase fruit and vegetable consumption to the nationally recommended 5 per day. Interventions that produce effects of this magnitude are rare—only 2% of all the interventions in Lipsey and Wilson's study (2001) achieved this effect size. The identification of successful interventions, together with insights into which intervention components might be attended to in the future, should enable

---

* The standardized mean difference, which is calculated as the difference in means between control and intervention groups divided by the population standard deviation.

future interventions to concentrate on approaches that work (or might work). The evidence-base does suggest that it is possible to increase fruit and vegetable consumption toward recommended levels but also that existing strategies will need some development if this goal is to be achieved.

In an earlier review, Hursti and Sjödén (1997) noted that, because even modest effects may be beneficial for cardiovascular disease if maintained over long periods, research should concentrate on how the effects on an intervention program might be maintained. In this regard, even the small effect sizes found in most of the studies may have a clinical significance that is not apparent at first glance. Indeed, the way in which two of the interventions (Baranowski et al., 2000; Reynolds et al., 2000) succeeded in sustaining their increased consumption over a year could be worthy of further examination. It is worth noting here that some of the strategies employed in these interventions were also supported by the findings of the children's views synthesis (discussed later).

Can Interventions Persuade Children to Try New Fruits or Vegetables?

Three studies evaluated the success of interventions in attempting to persuade children to try new fruits and/or vegetables: Hendy (1999), Shannon et al. (1982) and Wardle et al. (2003a). The interventions in both Hendy (1999) and Wardle et al. (2003a) contained elements of modeling — of children observing an adult eating the targeted foods before consuming it themselves. Both studies also compared different types of intervention: in Hendy's case the comparison was between different teacher actions — simple exposure, modeling, rewards, insisting that children try one bite, and choice-offering. Wardle et al. (2003a) compared two of the same techniques: reward and exposure. Both reports cite "Over Justification Theory"* in suggesting that, in the longer term, rewards are counterproductive and devalue the food in the eyes of the child (although there might be some short-term effect).

RESULTS: "QUALITATIVE" SYNTHESIS OF "VIEWS" STUDIES

The findings and conclusions of each study were copied verbatim as reported by study authors into our software (Thomas, 2002). This software enabled us to record answers to a large number of questions about each

---

* This theory suggests that rewards decrease personal motivation over time since a person's behavior can become dependent on receiving a reward.

study. Some of the most important questions asked reviewers to group findings according to their ability to illuminate the following:

- What are children's perceptions of and attitudes toward healthy eating?
- What do children think stops them from eating healthily?
- What do children think helps them to eat healthily?
- What ideas do children have for what could or should be done to promote their healthy eating?

The study findings and conclusions within each of these groups were exported to NVivo (Version 2.0) from QSR Software, a specialist software package for undertaking qualitative analysis of textual data.

Three of the review authors carried out the synthesis, meeting on a total of six occasions (for periods of between 2 and 5 hours) over a 3-week period. Synthesis methods broadly followed guidelines for the thematic analysis of textual data collected in the context of primary research. In this case, the textual data that we used in our analysis were the study authors' descriptions of their findings. This is a relatively new method of bringing the results of these types of studies together. We utilized authors' accounts of their results rather than the conclusions that they had drawn from them.

The synthesis of findings from the views studies was generated from studies involving 1091 children and 92 mothers living in Scotland or England. Although reporting limitations make it difficult to establish precisely, these findings appear to have been derived from children from both lower and higher socioeconomic status families. Due to a near complete lack of reporting on ethnicity, it is not clear to what extent the findings reported here were derived from, or are applicable to, children from ethnic minority groups.

Although we did not exclude any studies on the grounds of quality, the reader should be aware that not all studies contributing findings to the synthesis were of the same methodological standard. In particular, three studies were judged to be of a poorer quality, meeting 6 or less of the 12 criteria that we used to assess the quality of these studies (Edwards and Hartwell, 2002; Neale et al., 1998; Tilston et al., 1991). The results of our quality assessment would have provided a good rationale for not allowing the findings of these three studies to contribute their findings to our overall synthesis of children's views. We decided not to exclude these studies, however, because when we checked, their findings did not contradict those from studies of a higher quality. In fact, these studies had very little to contribute to the synthesis.

We examined the findings of each study in turn, with every sentence or paragraph within the report of the findings assigned a code to describe it (e.g., children prefer fruit to vegetables). This process created a total of

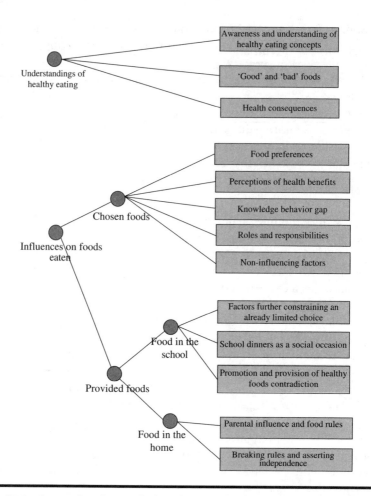

Figure 15.3 Interrelated Descriptive Themes Identified Across Studies of Children's Views (*n* = 8).

36 initial codes. Reviewers looked for similarities and differences between the codes in order to start grouping them. New codes were created to capture the meaning of groups of initial codes. This process resulted in a structure with several layers to organize a total of 12 descriptive themes, which are presented in Figure 15.3.

In an iterative process, barriers to, and facilitators of, healthy eating and implications for intervention development were *inferred* from the descriptive themes. This was a cyclical process that was repeated until the themes were sufficiently abstract to encompass all of our initial descriptive themes. The *analytical* themes that emerged as a result of this process and their associated barriers, facilitators, and implications for interventions are shown in Table 15.1.

Table 15.1 Barriers, Facilitators, and Implications for Interventions to Promote Increased Fruit and Vegetable Intake Among Children

Themes from Children's Views	Barriers	Facilitators	Implications for Intervention Development
Children do not see it as their role to be interested in health	Children dismiss the health consequences of eating or not eating healthily and prioritize taste preferences. They consider taste, not health, to be a key influence on food choice Food labeled as healthy may lead children to reject them ("I don't like them so they must be healthy") Children do not see buying healthy foods as a legitimate use of their money		Brand fruit and vegetables as "tasty" rather than "healthy" Promote children's favorite fruit and vegetables or target the ones they do not like the taste of
Children do not see future health consequences as personally relevant or credible	Children dismiss possible health consequences of not eating healthily for them personally ("don't care") Children feel that health messages (e.g., "sweets rot your teeth") do not match their actual experience	Immediate health consequences may be more relevant to children (e.g., effects on skin; energy to move around)	Reduce emphasis on health messages, particularly those that concern future health Make health messages credible and relevant for children (e.g., a celebrity promoting fruit)

(continued)

Table 15.1   Barriers, Facilitators, and Implications for Interventions to Promote Increased Fruit and Vegetable Intake Among Children (Continued)

Themes from Children's Views	Barriers	Facilitators	Implications for Intervention Development
Fruit, vegetables, and confectionary have very different meanings for children	Children do not like (some) vegetables because they taste sour ("yucky") Children do not like large and hard vegetables Eating sweets is a social and 'exciting' activity to be shared with friends and siblings	Children like fruit because it is sweet Fruit is preferred to vegetables and is liked almost as much as confectionary Children prefer brightly colored, small, soft, juicy and sweet vegetables	Do not promote fruit and vegetables in the same way; do not promote fruit and vegetables within the same intervention; if fruit and vegetables are promoted in the same intervention, treat them differently Brand fruit and vegetables as an "exciting" or child-relevant product, as well as a "tasty" one
Children actively seek ways to exercise their own choices with regard to foods	Eating sweets, despite parental rules, is a way for children to assert their own independence Children can feel under pressure to choose and eat food quickly in school	For girls, choosing healthy foods appears to be a way for them to exercise their own choice	Create situations for children to have ownership over their food choices, e.g., make eating fruit a choice not an obligation
Children value eating as a social occasion	Eating sweets is a social and 'exciting' activity to be shared with friends and siblings	Children like to sit with their friends at school	Brand fruit and vegetables as an 'exciting' or child-relevant product, as well as a 'tasty' one Create situations for children to have ownership over the social context in which they eat their food

Table 15.1 Barriers, Facilitators, and Implications for Interventions to Promote Increased Fruit and Vegetable Intake Among Children (Continued)

Themes from Children's Views	Barriers	Facilitators	Implications for Intervention Development
Children recognize the contradiction between what 'adults' promote in theory and what is provided in practice	Easy access to tempting (unhealthy) foods Contradiction between the promotion of healthy foods in the classroom and the provision of unhealthy foods in the school dining hall		Ensure that messages promoting fruit and vegetables are supported by appropriate access to fruit and vegetables

Analytical Themes and Barriers and Facilitators

The barriers and facilitators framework of our review did not appear to match the way children included in the studies talked about healthy eating. Children did not identify readily (and do not appear to have been asked about directly) things that helped them to eat, or stopped them from eating, healthily. It was, however, possible for reviewers to interpret, from the perspectives that children did express, the issues articulated by children that might act as barriers or facilitators for healthy eating and for increasing their consumption of fruits and vegetables in particular.

Children Do not See it as Their Role to be Interested in Health (Theme 1)

Children identified readily that taste was the major concern for them when selecting food and that health was either a secondary factor or, in some cases, a reason for rejecting food. This idea—that children were not motivated by health-related incentives—was further exemplified when children stated that they did not see buying healthy food as being a legitimate use of their money.

The reviewers felt that these barriers could be addressed by a shift in emphasis on the presentation of healthy foods. As one child noted astutely, "All adverts for healthy stuff go on about healthy things. The adverts for unhealthy things tell you how nice they taste" (Dixey et al., 2001, p. 75). This idea, coupled with children's identification of taste as the most important factor for them when choosing food, indicates that branding fruits and vegetables as a "tasty," "exciting," or child-relevant product could be far more effective in increasing consumption than a focus on health messages.

The issue of presentation seemed to be a salient one for children. One recommendation that came directly from a child was the suggestion that fresh fruit could be advertised with a famous celebrity. The children in the study by Dixey et al. (2001) discussed food and advertising and acknowledged that they liked the advertisements involving football celebrities, although these advertisements involved unhealthy food. The child's recommendation and the views on advertising suggest that children would be receptive to advertisements that link fruit with their interests.

Furthermore, although the children in the study by Gibson et al., rated health as the second most important factor in choosing food, it was reported by other children that they actively avoided foods they knew to be healthy (Dixey et al., 2001). The views presented thus indicate that health promotion drives would do well to promote "tastiness" as the primary appeal of fruit and vegetables with health as a secondary, if not invisible, factor.

A second recommendation for future practice also stemmed from children's views about taste. Children indicated that certain vegetables were preferable to others. Larger vegetables such as cauliflower, cabbage, turnip, and vegetables with a sour taste were identified as being not very appealing to children. Smaller, sweeter vegetables, such as peas, sweet corn, and baked beans were reported to be much more appealing. Therefore, it would seem appropriate for products and interventions to link in to what children are saying about the vegetables that they prefer.

Future Health Consequences (Theme 2)

Data from a number of studies indicated that health messages about the future consequences of not eating healthily are not seen as important for children. Indeed, children suggested that these messages are being received, but they are unconcerned about such issues.

Furthermore, children found that the health messages they were receiving did not stand up to their scrutiny. Children demonstrated an awareness of health messages, but their own experience told them that not eating healthily did not seem to have a negative impact on them, thereby negating the credibility of the messages they received. For example, the article by Mauthner et al. (1993; p. 28) cites a child whose experience tells her that "I like sweets. When I eat it doesn't wobble my teeth."

Furthermore, messages about cancer or heart disease are even less likely than tooth decay to match a child's personal health experiences and are, therefore, even less suitable as messages for children.

Children's discussions of their understanding of healthy diets revealed a further issue regarding health messages. In children's accounts of what constitutes a healthy diet, the focus was overwhelmingly on not having too much fat in the diet and to a lesser extent the impact of sugar on dental health. Some children in the Tilston study (Tilston et al., 1991, p. 28) even indicated an awareness of the link between salt and blood pressure. Although some children indicated an understanding of balance, they did not highlight fruit and vegetables as constituting part of a healthy diet. Therefore, it would seem that children are able to take health messages on board, and it may be worth giving some clear messages about fruit and vegetable intake to counterbalance the overwhelming viewpoint that health messages are primarily about fat. However, as noted earlier, children's views indicated that health messages are not of interest to them; therefore, if health messages are to be conveyed, the children's views presented here suggested that these messages should be child relevant and given less importance than messages about taste.

Fruit, Vegetables, and Confectionary (Theme 3)

Children view fruit and vegetables as very different in nature. The children in several studies indicated that they liked fruit (e.g., Gibson et al., 1998; Neale et al., 1998; Turner et al., 1995). The children in the Gibson et al., study (1998) rated fruit nearly as highly as confectionery, whereas in the study by Mauthner et al. (1993) children said that they like fruit because it "tastes sweet." No children in any study indicated that they disliked fruit. However, there was much evidence to suggest that vegetables were less well liked. In the study by Gibson et al. the children's rated liking for vegetables was significantly lower than their rating for fruit. The Edwards and Hartwell (2002) study also found that children rated acceptability of fruit higher than acceptability of vegetables. This would suggest that fruits and vegetables should not be promoted in the same way.

Such findings must be considered in association with the above findings on barriers and facilitators, i.e., that sweet taste is an important facilitator for children and that links to health benefits can be a barrier; "everything that is healthy tastes awful." Therefore, presenting a health message linking vegetables and fruit may in fact be detrimental to fruit consumption, as fruit would move from a category associated with "tasty" food to a category associated with "healthy" and by definition "awful" food.

Children Exercise Choices (Theme 4)

Children often expressed views about the importance, for them, of exercising choice. For example, eating sweets, despite parental rules, is a way for children to assert their own independence. Some children in the study by Mauthner et al. (1993) emphasized that they felt unhappy when they were not able to exercise their choice fully, such as feeling under pressure to choose and eat food quickly in school. Girls stated that choosing healthy foods was a way for them to exercise their own choice.

These findings point to the importance of healthy eating interventions enabling children to make healthy choices and to have ownership over their food choices. For example, interventions could involve children in the planning of healthy school menus (see Shepherd et al., 2001; Thomas et al., 2003).

Eating as a Social Occasion (Theme 5)

Many comments made by children showed that they valued eating for its social aspect. This is relevant to the development of appropriate interventions. Some children suggested that eating sweets is a social and "exciting" activity to be shared with friends and siblings. If fruit products could be branded and presented as something child relevant they would have the potential for being valued as part of a similarly exciting activity. Other children discussed the context in which they eat in terms of the social aspect of the school lunch. The fact that children said that they like to sit with their friends at school suggests that healthy eating interventions should foster the social side of lunchtime to fuse healthy eating with an enjoyable occasion.

Contradictions in Theory and Practice (Theme 6)

The children themselves provided three recommendations for the promotion of fruit and vegetable consumption. The recommendations came from the study by Dixey et al. (2001) in which children had been exposed to the "APPLES Project" intervention. One recommendation was that the fruit snack shops, which were part of the "APPLES Project," would be worth running permanently. This suggestion ties in with their views about having access to healthy foods being a facilitator and being able to exercise choice. It may also be worth involving children in the choice of selection of fruits available.

Some children observed that, even when they did take the health promotion messages on board, good intentions were often defeated by the provision of unhealthy foods by others or ease of access to unhealthy foods, which tempted them when making their own choice. To this end, it would be important for interventions to ensure that messages promoting fruit and vegetables are supported by appropriate access to them.

RESULTS: INTEGRATING "QUANTITATIVE" AND "QUALITATIVE" SYNTHESES

A methodological and conceptual matrix developed in earlier reviews (Harden et al., 2001; Rees et al., 2001; Shepherd et al., 2001) was used to juxtapose the findings of views studies against the findings of "trials."

Three questions guided the cross-study synthesis:

1. Which interventions promoting an increase in children's consumption of fruit and vegetables match recommendations derived from children's views and experiences of healthy eating?
2. Do those interventions that match children's views show bigger effect sizes in their evaluations and/or explain heterogeneity between studies than those that do not?
3. Which recommendations derived from children's views have yet to be addressed by interventions evaluated by trials?

The products of the synthesis of the findings of children's views studies (the implications for interventions organized by analytical theme and associated barriers and facilitators) were used as the starting point for the cross-study synthesis. These are listed in the left-hand column of the conceptual and methodological matrix shown in Table 15.2. Each intervention implication from the "views" studies was compared with the interventions tested by the trials to see whether the intervention included a component that matched what the children had said might be effective. Matching interventions were sought from our pool of high- or medium-quality trials first of all. If no or few matches were found, matching interventions were sought from our pool of other trials of a lower methodological quality. Matches and gaps are noted in the right hand columns of the matrix (Table 15.2).

Of those interventions with components matching children's views, some were clearly effective, some were unclear in their effects, but none was harmful. Those interventions that led to bigger increases in fruits and/or vegetables consumed included one or more of the following components that matched children's views: the promotion of fruit and vegetables in separate interventions or in different ways within the same intervention; a reduction in the emphasis on health messages; or the promotion of fruit and vegetables in educational materials accompanied by access to fruit and vegetables. However, the effectiveness of interventions with the following components matching children's views is unclear, and further evaluation is required: branding fruit and vegetables as exciting or child-relevant products and encouraging situations for children to express choice.

Gaps between evaluated interventions and children's views revealed the following opportunities for developing and evaluating innovative interventions based on children's views (Table 15.2). These included: branding fruit and vegetables as "tasty" rather than "healthy," creating opportunities for children to influence the social context in which they eat, and making health messages credible for children.

Table 15.2   Numbers of Reliable and Other Trials Addressing Implications for Interventions Identified from Children's Views

Implication	Reliable Trials Addressing Implication	Other Trials Addressing Implication
(1) Brand fruit and vegetables as a "tasty" rather than a "healthy" product	None identified	None identified
(2) Promote children's favorite fruit and vegetables or target the ones they do not like	Hendy (1999) Wardle et al. (2003c) Wardle et al. (2003d)	None identified
(3) Reduce emphasis on health messages, particularly those which concern future health	Hendy (1999) Liquori et al. (1998) Smolak et al. (1998) Wardle et al. (2003a) Wardle et al. (2003b)	Domel et al. (1993)
(4) Make health messages credible and relevant for children	None identified	None identified
(5a) Do not promote fruit and vegetables in the same intervention	Liquori et al. (1998) Moore (2001) Wardle et al. (2003a) Wardle et al. (2003b)	Smith and Justice (1979)
(5b) If promoting fruit and vegetables in the same intervention, treat them differently	Baranowski et al. (2000)	None identified
(6) Brand fruit and vegetables as an "exciting" or child-relevant product, as well as a tasty one	Anderson et al. (2000) Baranowski et al. (2000) Hopper et al. (1996) Perry et al. (1998a) Reynolds et al. (2000) Shannon et al. (1982)	Boaz et al. (1998) Domel et al. (1993) Foerster et al. (1998) Friel et al. (1998) Lawatsch (1990) Lowe et al. (submitted for publication)
(7) Create situations for children to have ownership over their food choices	Moore (2001)	Foerster et al. (1998)
(8) Create opportunities for children to have ownership over the social context in which they eat their food	None identified	None identified

Table 15.2 Numbers of Reliable and Other Trials Addressing Implications for Interventions Identified from Children's Views (Continued)

Implication	Reliable Trials Addressing Implication	Other Trials Addressing Implication
(9) Ensure that messages promoting fruit and vegetables are supported by appropriate access to fruit and vegetables	Anderson et al. (2000) Parcel et al. (1999) Perry et al. (1998a) Perry et al. (1998b) Reynolds et al. (2000)	Resnicow et al. (1992)

DISCUSSION

This is the first review of which we are aware that attempts to analyze and synthesize, in a systematic way, the findings from studies of children's views and experiences of food and healthy eating and tries to integrate these findings with those derived from effectiveness studies.

Discussion: "Quantitative" Synthesis of Trials

Pooling the findings of 19 studies revealed that, on average, interventions are able to increase children's fruit intake by one fifth of a portion a day and their vegetable intake by a little less than one fifth of a portion a day. However, further analysis revealed that the effects of interventions that focused more specifically on healthy eating were nearly three times greater than those that tried to target healthy eating alongside physical activity and/or smoking. These increases look negligible, however, as discussed above; the effect sizes attained by the studies in this review were only a little less than the average across a number of different fields. Thus, judging from the experiences of a large number of other studies, only an exceptionally successful intervention would be able to increase fruit and vegetable consumption to the nationally recommended five per day. Interventions that produce effects of this magnitude are rare. The evidence base does suggest that it is possible to increase fruit and vegetable consumption toward recommended levels but also that existing strategies will need some development if this goal is to be achieved.

An interesting finding of the effectiveness synthesis was that it appears to be easier to increase children's consumption of fruits than vegetables. A small number of studies attempted to address children's apparent greater dislike for vegetables by "exposing them" to new or previously disliked vegetables. Their results revealed that it is possible to get children to try these vegetables, but it is unclear whether such strategies would lead to increases in children's everyday consumption.

We identified 10 other systematic reviews that included studies evaluating interventions for promoting healthy eating among children, although none of these had exactly the same scope or population focus as this review. These reviews also found small effects of interventions. For example, McArthur (1998) entered 12 studies of interventions to encourage "heart healthy" eating behaviors of children aged 9 to 11 years old into a meta-analysis and found a pooled effect size of 0.24. Hursti and Sjödén (1997, p. 110) concluded that, "Most of the reported changes in dietary behavior were modest" in their review of interventions to change food habits among children. Similarly, Contento and colleagues (1992, p. 257) concluded that the impact of general nutrition education on behavior was "minimal" in their review of programs among school-aged children.

Two reviews had findings different from ours regarding the lesser effectiveness of interventions targeting physical activity as well as healthy eating for increasing children's fruit and vegetable intake. A key conclusion from the review undertaken by the Centre for Reviews and Dissemination (NHS Centre for Reviews and Dissemination, 2002) was that programs that "promote physical activity, the modification of dietary intake and the targeting of sedentary behaviors" have been found to be effective. Resnicow and Robinson (1997) drew similar conclusions in their review. These apparently contradictory conclusions may simply be due to the differing scope of these reviews from ours: obesity prevention and treatment (NHS Centre for Reviews and Dissemination, 1997) and cardiovascular disease prevention (Resnicow and Robinson, 1997). Both of these reviews also extended beyond studies with children. The review by Ciliska and colleagues (2000) found that the most effective interventions in their review on promoting fruit and vegetable consumption gave clear and undiluted messages about fruit and vegetables — a finding that is supported by this review.

Discussion: "Qualitative" Synthesis of "Views" Studies

The views from children provided valuable insights into their experiences of food and healthy eating. Our synthesis revealed a number of contextual issues that any program that aims to promote healthy eating among children should consider. For example, children do not see it as their role to be interested in health and they do not see messages regarding possible future health consequences as being relevant or credible. Promoting fruits and vegetables on health grounds, therefore, may have little currency among children. Children identified the "here-and-now" aspects of eating food as important (e.g., children valued eating as a social occasion), and different foods had different meanings for them. Eating certain foods such as confectionary was seen as risky, exciting, and as a way of breaking adult rules and asserting independence. Given that it is not usual for children to have a choice about what they eat, they talked of seeking ways actively to exercise their own choices (e.g., throwing away the "healthy foods" that parents had provided for them). Children recognized readily the contradiction between what "adults" promote in theory and what is provided in practice. For instance, the school cafeteria often came under fire for not providing the kinds of healthy foods that adults promote.

With this synthesis of children's views studies, we constructed a broad picture of the issues that children from diverse backgrounds regard as important. Some differences between children were found in several studies in relation to sociodemographic factors, such as sex, age, and socioeconomic status. However, the differences explored in the studies tended to be studied solely in relation to knowledge and behaviors rather than children's understandings and experiences. This represents a gap for future research.

We are aware of no other systematic review that has attempted to synthesize the findings of this type of study. The picture that emerges from our review highlights a number of deeper issues concerning the construction of children and childhood that resonate with other primary research attempts to understand health and social issues from children's perspectives (Mayall et al., 1996; O'Brien et al., 2000; Scott, 2000). Many of the children's insights into food and healthy eating may be similar to those that might be found among adults. Our review has found, however, that professionals who design and evaluate interventions do not always consider these insights.

Discussion: Integrating "Quantitative" and "Qualitative" Syntheses

The findings of our views synthesis suggested 10 implications for the development of appropriate interventions (see Table 15.1). Comparison of evaluated interventions to these implications in our cross-study synthesis revealed matches, mismatches, and gaps (see Table 15.2). Furthermore, we found evidence to suggest that some interventions that matched the implications derived from children's views led to bigger effects than those that did not. Interventions that did the following resulted in bigger increases in fruits and/or vegetables consumed: those that promoted fruits and vegetables separately or in different ways, those that reduced or removed any emphasis on health messages, and those that supported the promotion of fruits and vegetables in educational materials with access to fruits and vegetables. However, we found no difference in effects when comparisons were made between interventions that promoted fruits and vegetables in ostensibly exciting and child-relevant ways and those that did not. It was not clear why this might be so, and further research is warranted to examine children's views on whether/how fruits and vegetables could be made more exciting to them.

Implications for Policy, Practice, and Research

In an earlier review (discussed above), Hursti and Sjödén (1997) noted that, because even modest effects may be beneficial for cardiovascular disease if maintained over long periods, research should concentrate on how the effects of an intervention program might be maintained. In this regard, even the small effect sizes found in most of the studies may have a clinical significance that is not apparent at first glance. Indeed, the way in which two of the interventions (Baranowski et al., 2000; Reynolds et al., 2000) succeeded in sustaining their increased consumption over a year could be worthy of further examination. It is worth noting here that some of the strategies employed in these interventions were also supported by the findings of the children's views synthesis.

The evidence from the cross-study synthesis, that higher effect sizes were correlated with some of the implications from children's views, further supports the need to explore areas that children suggest are important but that have not been researched extensively. Interventions that matched two of the implications derived from children's views (to reduce the emphasis on health messages or to promote fruits and vegetables differently or separately) had higher effect sizes than studies that did not. This finding suggests key messages for future policy, practice, and research. For policy and practice, the current messages that promote fruits and vegetables together, such as the "five-a-day" messages promoted by organizations including the U.K. Department of Health, may not be suitable for children and that practitioners should be focusing on promoting fruits and vegetables separately, or differently, as in the government's proposed U.K. National School Fruit Scheme rolled out in 2003.

A second message for policy would be that in order to increase fruit and vegetable intake among children, a move away from current health messages about fruits and vegetables might be necessary. Again one of the two key aims of the U.K. Department of Health's "5 A DAY" initiative is to raise awareness of health benefits, which may be suitable for adults; however, further work may have to be undertaken in order for this to be successful with children.

The implications for research are, first, to ensure that all the implications arising from children's views are considered for their potential for informing intervention design; second, that children's views should be incorporated in the development and evaluation of any future healthy eating initiatives. For instance, there is an opportunity for intervention design to address the creation of situations that allow children to have ownership over their food choices and over the social context in which they eat their food. As Roberts (2000) discussed, the involvement of the recipients of interventions in their development is a matter of research ethics; it is not sufficient simply to listen to children, it is also important to act upon their views. However, the findings of our review suggest that incorporating children's views into the development of interventions is not only desirable from an ethical viewpoint but is also necessary in order to develop effective interventions.

Another area suggested by the Cross-study synthesis, which would be worth exploring in future developments of potentially effective interventions, relates to those qualitatives children's views that were not found to have any impact on existing outcome evaluations. In particular, the sixth implication from Table 15.2, that fruit and vegetables should be branded as an exciting or child-relevant product, may need further unpacking. Although several studies addressed this issue, they may not have done so in "exciting" or relevant ways. For example, interventions employing cartoon characters, stickers, and games were classified for this review as using methods to make fruit and vegetables "exciting" and relevant; however, children suggested using television advertising campaigns featuring famous "celebrities" such as football stars. For example, the "Food Dudes" intervention being rolled out in Scotland (Forth Valley NHS Board, 2003) using "cartoon" techniques similar to those in the studies included in the subgroup analysis in the cross-study synthesis; suggest that further research with children into the intervention components that they would find "exciting" or relevant could be valuable.

The principle that practice should involve children and focus on what they are saying is an important one. But it must be recognized that not all of the implications will be easy to incorporate into practice. For example, if children are using their own experience of being healthy despite eating unhealthily and choosing to disregard messages about long-term health, it may not be possible to develop messages that are credible or relevant for them. Practitioners might choose to prioritize those messages from children that are less problematic, for example, promoting fruit and vegetables as being "tasty" rather than healthy (an implication for which no matching interventions were found).

CONCLUSIONS

Our review has uncovered a relatively solid evidence-base for informing policy and practice for the promotion of fruits and vegetables to children aged 4 to 10. Pooling the findings from good-quality trials indicated that interventions can have a small but significant positive effect, increasing children's fruit intake by one fifth of a portion per day and their vegetable intake by nearly one fifth of a portion per day. Assessing the significance of these effects requires their translation into estimates of health gain and clinical significance together with their potential savings for health care services. Our synthesis of effectiveness research has indicated the types of interventions and their components that lead to larger or smaller effects (e.g., those that targeted families at high risk for cardiovascular disease had higher effect sizes; those interventions that diluted their focus on fruit and vegetables by also aiming to promote physical activity or other forms of healthy eating tended to have lower effect sizes).

Clear implications regarding the development of appropriate interventions were derived from studies eliciting children's own perspectives on food, eating, and healthy eating. Moreover, within our cross-study synthesis, we found a relationship between what children say is important and the intervention's effectiveness. We were able to use these findings to identify further intervention components that lead to larger effects (e.g., interventions that did not emphasize the health benefits of fruits and vegetables showed larger effects than those that did).

Our cross-study synthesis also highlighted a number of promising directions for the future development and testing of interventions to promote fruits and vegetables. In particular, there is scope to explore the effect of interventions that brand fruits and vegetables as being "tasty" rather than "healthy" and in creating opportunities for children to influence the social context in which they eat. Additionally, one challenging implication calls for health messages to be made relevant and credible for children. The studies of children's views have demonstrated that the perspectives of those people who are the "targets" of public health interventions can be a valuable source of knowledge. We recommend, therefore, that relevant stakeholders — researchers, practitioners, children, and their parents—should work in partnership to develop future interventions.

REFERENCES

Anderson, A.S., Hetherington, M., Adamson, A., Porteous, L., Higgins, C., Foster, E., Stead, M., and Ha, M., *The Development and Evaluation of a Novel School-Based Intervention to Increase Fruit and Vegetable Intake in Children,* Food Standards Agency, London, 2000.

Auld, G.W., Romaniello, C., Heimendinger, J., Hambidge, C., and Hambidge, M., Outcomes from a school-based nutrition education program alternating special resource teachers and classroom teachers, *J. School Health,* 69, 403–408, 1999.

Baranowski, T., Davis, M., Resnicow, K., Baranowski, J., Doyle, C., Lin, L.S., Smith, M., and Wang, D.T., Gimme 5 fruit, juice, and vegetables for fun and health: outcome evaluation, *Health Educ. Behav.,* 27, 96–111, 2000.

Boaz, A., Ziebland, S., Wyke, S., and Walker, J., A "five-a-day" fruit and vegetable pack for primary school children. Part II: controlled evaluation in two Scottish schools, *Health Educ. J.* 57, 105–116, 1998.

Brannen, J., Dodd, K., Oakley, A., and Storey, P., *Young People, Health and Family Life,* Open University Press, Buckingham, 1994.

Centre for Reviews and Dissemination, *The Prevention and Treatment of Childhood Obesity,* Centre for Reviews and Dissemination, York, 2002.

Ciliska, D., Miles, E., O'Brien, M.A., Turl, C., Tomasik, H.H., Donovan, U., and Beyers, J., Effectiveness of community-based interventions to increase fruit and vegetable consumption, *J. Nutr. Educ.*, 32, 341–352, 2000.

Contento, I.R., Manning, A.D., and Shannon, B., Research perspective on school-based nutrition education, *J. Nutr. Educ.*, 24, 247–260, 1992.

Cooper, H. and Hedges, L.V., *The Handbook of Research Synthesis*, Russell Sage Foundation, New York, 1994.

Department of Health, *National Diet and Nutrition Survey: Young People Aged 4–18*, The Stationery Office, London, 2000.

Dixey, R., Sahota, P., Atwal, S., and Turner, A., Children talking about healthy eating: data from focus groups with 300 9–11-year-olds, *Nutr. Bull.*, 26, 71–79, 2001.

Domel, S.B., Baranowski, T., Davis, H., Thompson, W.O., Leonard, S.B., Riley, P., Baranowski, J., Dudovitz, B., and Smyth, M., Development and evaluation of a school intervention to increase fruit and vegetable consumption among 4th and 5th grade students, *J. Nutr. Educ.*, 25, 345–349, 1993.

Doyle, M. and Hosfield, N., *Health Survey for England 2001: Fruit and Vegetable Consumption*, HMSO, London, 2003.

Edwards, J.S.A. and Hartwell, H.H., Fruit and vegetables — attitudes and knowledge of primary school children, *J. Human Nutr. Diet.*, 15, 365–374, 2002.

Epstein, L.H., Gordy, C.C., Raynor, H.A., Beddome, M., Kilanowski, C.K., and Paluch, R., Increasing fruit and vegetable intake and decreasing fat and sugar intake in families at risk for childhood obesity, *Obes. Res.*, 9, 171–178, 2001.

Foerster, S.B., Gregson, J., Beall, D.L., Hudes, M., Magnuson, H., Livingston, S., Davis, M.A., Joy, A.B.J., and Garbolino T., The California children's 5 a Day Power Play! campaign: evaluation of a large-scale social marketing initiative, *Family Comm. Health*, 21, 46–64, 1998.

Forth Valley NHS Board, *Enter the Food Dudes — and Healthy Food Becomes Cool*, Press release, available at http://www.show.scot.nhs.uk/nhsfv/news/Media%20 Release%20-%20Food%20Dudes%206%20Jan%202003.doc, 6 January, 2003.

Friel, S., Kelleher, C., Campbell, P., and Nolan, G., Evaluation of the Nutrition Education at Primary School (NEAPS) programme, *Public Health Nutr.*, 2, 549–555, 1999.

Fulton, J.E., McGuire, M.T., Caspersen, C.J., Dietz, W.H., Interventions for weight loss and weight gain prevention among youth: current issues, *Sports Med.*, 31, 153–165, 2001.

Gibson, E.L., Wardle, J., and Watts, C.J., Fruit and vegetable consumption, nutrition knowledge and beliefs in mothers and children, *Appetite*, 31, 205–228, 1998.

Harden A. and Thomas J., Methodological issues in combining diverse study types in systematic reviews, *Int. J. Soc. Res. Methodol.*, in press.

Harden, A., Rees, R., Shepherd, J., Brunton, G., Oliver, S., Oakley, A., *Young people and mental health: a systematic review of research on barriers and facilitators*, EPPI-Centre, Social Science Research Unit, Institute of Education, University of London, London, 2001.

Hendy, H.M., Comparison of five teacher actions to encourage children's new food acceptance, *Ann. Behav. Med.*, 21, 20–26, 1999.

Hopper, C.A., Munoz, K.D., Gruber, M.B., MacConnie, S., Schonfeldt, B., and Shunk, T., A school-based cardiovascular exercise and nutrition program with parent participation: an evaluation study, *Children's Health Care*, 25, 221–235, 1996.

Hursti, U.K. and Sjödén, P., Changing food habits in children and adolescents: experiences from intervention studies, *Scand. J. Nutr.*, 41, 102–110, 1997.

Lawatsch, D., A comparison of two teaching strategies on nutrition knowledge, attitudes, and food behavior of pre-school children, *J. Nutr. Educ.*, 22, 117, 1990.

Lipsey, M.W. and Wilson, D.B., *Practical Meta-Analysis*, Sage Publications Inc., London, 2001.

Liquori, T., Koch, P.D., Contento, I.R., and Castle, J., The Cookshop Program: outcome evaluation of a nutrition education program linking lunchroom food experiences with classroom cooking experiences, *J. Nutr. Educ.*, 30, 302–313, 1998.

Lowe, C.F., Horne, P.J., Tapper, K., and Bowdery, M., Evaluation of a peer modelling and rewards based programme to increase primary school children's consumption of fruit and vegetables, submitted for publication.

Lytle, L.A., *Nutrition Education for School-aged Children: A review of research*, Food and Consumer Service (USDA), ED428061, Washington, D.C., 1994.

Mauthner, M., Mayall, B., and Turner, S., *Children and food at primary school*, Social Science Research Unit, Institute of Education, University of London, London, 1993.

Mayall, B., Bendelow, G., Barker, S., Storey, P., and Veltman, M., *Children's Health in Primary School*, Falmer Press, London, 1996.

McArthur, D.B., Heart healthy eating behaviors of children following a school-based intervention: a meta-analysis, *Issues Compr. Pediatr. Nurs.*, 21, 35–48, 1998.

Moore, H. and Kindness, L., Establishing a research agenda for the health and well being of children and young people in the context of health promotion, in *Promoting the Health of Children and Young People: Setting a Research Agenda*, Moore, H., Ed., Health Education Authority, London, 1998.

Moore, L., *Are Fruit Tuck Shops in Primary Schools Effective in Increasing Pupils Fruit Consumption? A Randomised Controlled Trial*, Cardiff School of Social Sciences, Cardiff, 2001.

Murray, D.M., *Design and Analysis of Group-randomized Trials*, Oxford University Press, Oxford, 1998.

Neale, R.J., Otte, S., and Tilston, C.H., Fruit: comparisons of attitudes, knowledge and preferences of primary school children in England and Germany, *Zeitschrift für Ernährungswissenschaft*, 37, 128–130, 1998.

NHS Centre for Reviews and Dissemination, The prevention and treatment of obesity, *Effective Health Care*, 3, 1–13, 1997.

NHS Centre for Reviews and Dissemination, The prevention and treatment of childhood obesity, *Effective Health Care*, 7, 1–11, 2002.

O'Brien, M., Rustin, M., Jones, D., Sloan, D., Children's independent spatial mobility in the urban public realm, *Childhood*, 7, 257–277, 2000.

Parcel, G.S., Simons-Morton, B.G., O'Hara, N.M., Baranowski, T., and Wilson, B., School promotion of healthful diet and physical activity: impact on learning outcomes and self-reported behavior, *Health Educ. Q.*, 16, 181–199, 1989.

Peersman, G., *A Descriptive Mapping of Health Promotion Studies in Young People*, EPPI-Centre, Social Science Research Unit, Institute of Education, University of London, London, 1996.

Peersman, G., Oliver, S., and Oakley, A., *EPPI-Centre Review Guidelines: Data Collections for the EPIC Database*, EPPI-Centre, Social Science Research Unit, Institute of Education, University of London, London, 1997.

Perry, C.L., Bishop, D.B., Taylor, G., Murray, D.M., Mays, R.W., Dudovitz, B.S., Smyth, M., and Story, M., Changing fruit and vegetable consumption among children: the 5-a-Day Power Plus program in St. Paul, Minnesota, *Am. J. Public Health*, 88, 603–609, 1998a.

Perry, C.L., Lytle, L.A., Feldman, H., Nicklas, T., Stone, E., Zive, M., Garceau, A., and Kelder, S.H., Effects of the Child and Adolescent Trial for Cardiovascular Health (CATCH) on fruit and vegetable intake, *J. Nutr. Educ.,* 30, 354–360, 1998b.

Rees R., Harden, A., Shepherd, J., Brunton, G., Oliver, S., and Oakley, A., *Young People and Physical Activity: A Systematic Review of Research on Barriers and Facilitators,* EPPI-Centre, Social Science Research Unit, Institute of Education, University of London, London, 2001.

Resnicow, K., Cohn, L., Reinhardt, J., Cross, D., Futterman, R., Kirschner, E., Wynder, E.L., and Allegrante, J.P., A three-year evaluation of the Know Your Body Program in inner-city schoolchildren, *Health Educ. Q.,* 19, 463–480, 1992.

Resnicow, K. and Robinson, T.N., School-based cardiovascular disease prevention studies: review and synthesis, *Ann. Epidemiol.,* 7, S14–S31, 1997.

Reynolds, K.D., Franklin, F.A., Binkley, D., Raczynski, J.M., Harrington, K.F., Kirk, K.A., and Person, S., Increasing the fruit and vegetable consumption of fourth-graders: results from the HHHigh 5 project, *Prev. Med.,* 30, 309–319, 2000.

Roberts, H., Listening to children: and hearing them, in *Research with Children: Perspectives and Practice,* Christensen, P. and James, A., Eds., Falmer Press, London, 2000.

Scott, S., The impact of risk and parental risk anxiety on the everyday worlds of children, in *Children 5–16, Research Briefing No. 19.* ESRC, 2000.

Shannon, B., Graves, K., and Hart, M., Food behavior of elementary school students after receiving nutrition education, *J. Am. Diet. Assn,* 81, 428–434, 1982.

Shepherd, J., Harden, A., Rees, R., Brunton, G., Garcia, J., Oliver, S., and Oakley, A., *Young People and Healthy Eating: A Systematic Review of Research on Barriers and Facilitators,* EPPI-Centre, Social Science Research Unit, Institute of Education, University of London, London, 2001.

Shucksmith, J. and Hendry, L., *Health Issues and Adolescents: Growing up and Speaking Out,* Routledge, London , 1998.

Smith, H.M. and Justice C.L., Effects of nutrition programs on third grade students, *J. Nutr. Educ.,* 11, 92–95, 1979.

Smolak, L., Levine, M.P., and Schermer, F., A controlled evaluation of an elementary school primary prevention program for eating problems, *J. Psychosom. Res.,* 44, 339–353, 1998.

Thomas, J., Harden, A., Oakley, A., Oliver, S., Sutcliffe, K., Rees, R., Brunton, G., and Kavanagh, J., Integrating qualitative research with trials in systematic reviews: an example from public health, *BMJ,* 328, 1010–1012, 2004.

Thomas, J., Sutcliffe, K., Harden, A., Oakley, A., Oliver, S., Rees, R., Brunton, G., and Kavanagh, J., *Children and Healthy Eating: A Systematic Review of Barriers and Facilitators,* EPPI-Centre, Social Science Research Unit, Institute of Education, University of London, London, 2003.

Thomas., J., *EPPI-Reviewer© 2.0 (Web edition), EPPI-Centre Software,* Social Science Research Unit, Institute of Education, London, 2002.

Tilston, C.H., Gregson, K., and Neale, R.J., Dietary awareness of primary schoolchildren, *Br. Food J.,* 93, 25–29, 1991.

Turner, S., Mayall, B., and Mauthner, M., One big rush: dinnertime at school, *Health Ed., J.,* 54, 18-2, 1995.

Wardle, J., Herrera, M.L., Cooke, L., and Gibson, E.L., Modifying children's food preferences: the effects of exposure and reward on acceptance of an unfamiliar vegetable, *Eur. J. Clin. Nutr.* 57, 341–348, 2003c.

Wardle, J., Cooke, L.J., Gibson, E.L., Sapochnik, M., Sheiham A., and Lawson, M., Increasing children's acceptance of vegetables: a randomized trial of guidance to parents, *Appetite,* 40, 155–162, 2003d.

White, M., Carlin, L., Rankin, J., and Adamson, A., *Effectiveness of Interventions to Promote Healthy Eating in People from Minority Ethnic Groups: A Review,* Health Education Authority, Effectiveness Review 12, London, 1998.

World Health Organization, *Diet, Nutrition and the Prevention of Chronic Diseases. Report of a Joint WHO/FAO Expert Consultation,* WHO, Geneva, 2003.

16

DOES FOOD PROMOTION INFLUENCE CHILDREN'S DIET? A REVIEW OF THE EVIDENCE

Laura McDermott, Martine Stead, and Gerard Hastings

CONTENTS

INTRODUCTION

Diet and body weight are complex phenomena that are determined by an interaction of social, cultural, physiological, behavioral, and environmental factors (Escobar, 1999). A combination of these factors may be contributing to the dramatic rise of obesity among young people. In recent years, there has been growing interest in the role of environmental determinants of diet and health (Crockett and Sims, 1995), such as the supply and availability of food products and the influence of friends and family on food choice, for example.

One such environmental influence that has come under increased scrutiny is the heavy marketing of foods to children. A recent WHO report identified commercial food marketing as a possible contributory factor to childhood obesity (WHO/FAO, 2003). However, there has traditionally been little consensus regarding the role, if any, of food promotion in shaping children's dietary choices. Efforts to establish the impact of commercial marketing activity on the health of children have often been hindered by the complexity and contested nature of the evidence in this area. In the absence of a consensus on the evidence, debate has often been heated.

This chapter describes the methodology and findings from a systematic review of research into the impact of food promotion on children's food knowledge, preferences, and behavior (Hastings et al., 2003). The research was commissioned by the U.K. Food Standards Agency in an attempt to clarify the influence of food promotion on children's dietary behavior. The research team comprised marketing, nutrition, economics, and public policy academics from four leading U.K. universities (Strathclyde, Oxford, York, and City), and the findings were published in September, 2003.

This research adds to the current obesity debate by measuring the role of food promotion in children's food-related behavior. It is worth noting that the review looked at only one piece of the obesity puzzle. It was not concerned with the various other possible influences on children's diet, such as sedentary lifestyles, computer technology, or other challenging policy areas thought to be influencing obesity trends.

RESEARCH QUESTIONS

The review had two main tasks. The first task was to systematically review the extent and nature of food promotion to children. To address this issue, the following questions were developed:

1. What promotional channels are being used to target children?
2. What food items are being promoted to children?
3. What are the principal creative strategies used to target children?

The second and fundamental task of the review was to investigate the *effect* of commercial food promotion on children. For this task, the following research questions were asked:

1. Is there a causal link between food promotion and children's food knowledge, preferences, and behavior?
2. If food promotion is shown to have an effect on children's food knowledge, preferences, and behavior, what is the extent of this influence relative to other factors?
3. In the studies that demonstrate an effect of food promotion on children's food knowledge, preferences, and behavior, does this effect impact total category sales, brand switching, or both?

REVIEW METHODS

Systematic procedures were used to search, gather, and evaluate evidence for the review. Although systematic review methods are widely used in clinical medicine to investigate the effectiveness of specific treatments (see www.cochrane.org), their application in social science is more recent, and, because the phenomena under investigation are more complex, use is more challenging.

The generic framework for undertaking systematic reviews was fairly easily adapted for this review. The systematic review process is summarized in Figure 16.1.

Preliminary Literature Search and Development of Review Protocol

A preliminary search of the academic literature was undertaken to provide a clearer indication of the nature and size of the evidence base. Filtered searches for existing reviews and primary studies were undertaken on a small sample of relevant databases, including ABI/INFORM and PsycINFO. This process informed the development of a detailed review protocol that clearly set out the research questions and the search strategy for the review.

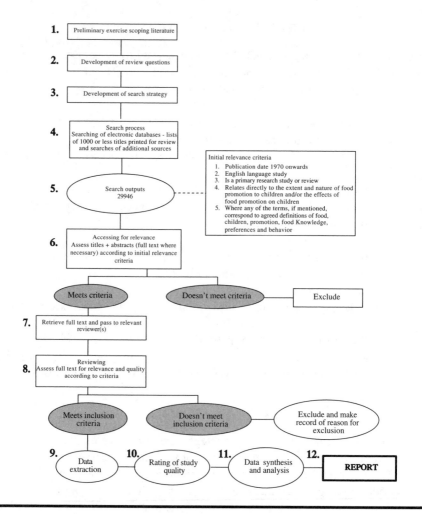

Figure 16.1   The Review Process.

The protocol is available on the U.K. Food Standards Agency's website (http://www.food.gov.uk/multimedia/pdfs/foodpromotiontochildren2.pdf).

In brief, three main methods were used to gather relevant data: an extensive search of electronic databases, searches of the "gray" literature, and personal contact with key people in the field. The primary source for relevant literature was electronic databases. Eleven databases (representative of relevant literatures such as psychology, marketing, and nutrition) were systematically searched using a combination of search terms, including food, food packaging, food advertising, food preferences, child(ren), youth, adolescents, marketing, advertising, promotion, and brands.

In addition, the reference lists from a sample of included studies were examined, and an "in-house" search for relevant literature was undertaken.

Together, these search methods yielded nearly 30,000 potentially relevant titles and abstracts. Full details of the search strategy (including a complete list of sources consulted, search terms used and searches undertaken) are provided in the published report (Hastings et al., 2003).

Relevance and Quality Assessment

Initial Relevance Assessment

All retrieved titles and abstracts were screened using predefined relevance criteria. Primary research studies or reviews published in the U.K. since 1970 were eligible for inclusion. Studies had to address directly the extent and nature of food promotion to children and/or its effects on their food knowledge, preferences, or behavior. Where mentioned, the terms *children, food, promotion,* or *knowledge, preferences,* and *behavior* had to correspond with agreed definitions developed specifically for the purpose of the review. For example, *food* was defined as all foods and nonalcoholic drinks, and *promotion* was defined as any form of commercial promotion, including advertising, branding, packaging, merchandising, and in-school marketing.

Two members of the review team independently undertook this first stage of relevance assessment, and a representative 10% sample of the reference lists was also independently reviewed for consistency.

Second-Stage Relevance and Quality Assessment

A total of 201 titles and abstracts met the initial relevance criteria, and each was retrieved in full text and then assessed against more stringent relevance and quality criteria. Studies relating to the extent and nature of food promotion to children, for example, had to adequately separate food from other items promoted to children to be considered relevant. Studies relating to the impact of food promotion to children, for example, had to measure children's exposure or response to *food* promotion as opposed to exposure or response to promotion in general. In terms of methodological quality, articles had to provide information about sample design, data collection methods, and data analysis procedures. At this early stage of assessment, all types of sample design (e.g., purposive, quota, and convenience) and study design (including experiments, surveys, observation, and qualitative methods) were permitted, provided that they were clearly described. The relative merits and limitations of these different designs were assessed in more depth at a later stage (see "Quality rating" below).

One hundred and one studies passed this second stage of relevance and quality assessment and were included in the review. Data extraction sheets, which provided a full but concise description of each study in terms of design, sample, methods, and procedures, analysis, and results, were completed for all included studies.

Quality Rating

The studies were too heterogeneous in terms of exposure type, subjects, settings, and outcomes to permit meta-analysis; a qualitative synthesis was therefore conducted.

Included studies were subject to a quality rating to help assess which studies' findings should be given more weight in drawing conclusions from the evidence. Studies were categorized, on the basis of their rating scores, as higher, medium, or lower quality. For studies examining the extent and nature of food promotion to children, quality criteria included the sample size, diversity and timing, thoroughness of the analysis, and clarity and completeness of the data reporting. For studies examining the effects of food promotion to children, quality criteria included the quality of the exposure measure, the quality of the effect(s) measure(s), the appropriateness of the analysis procedures, the extent and thoroughness of the analysis, and the clarity of and completeness of data reporting.

REVIEW FINDINGS

The Extent and Nature of Food Promotion to Children

What Promotional Channels are Being Used to Target Children?

The review showed that food is promoted to children more than any other product, with toys being the sole rival and then only at Christmas. Television is the principal medium for this advertising. Beyond television advertising, below-the-line promotional techniques such as sponsorship, in-school marketing, point-of-sale (i.e., promotional materials situated where consumers physically make a purchase), free samples of food items, free gifts/tokens (premiums) with food items, loyalty schemes, interactive food, novel packaging, tie-ins with movies, tie-ins with computer software, and other forms of wider brand building were also used.

There is some evidence that the dominance of television has recently begun to wane. The importance of strong, global branding reinforces a need for multifaceted communications combining television with merchandising, "tie-ins," and point-of-sale activity.

What Food Items are Being Promoted to Children?

Most studies undertook content analyses in order to determine what types of foods were promoted to children. Studies either looked at the relative amounts of advertising for specific foods, attempted to estimate their actual nutritional content, or made comparisons between the "advertised diet" and a defined recommended diet.

The review found that children's food advertising is dominated by the "big five" food items: presugared breakfast cereals, confectionary, savory snacks, soft drinks, and, in the last 10 years, fast-food restaurants. The advertised diet contrasts sharply with that recommended by public health guidelines — it is consistently higher in salt, sugar, and fat.

What Are the Principal Creative Strategies Used to Target Children?

Content analysis studies were examined to identify the creative strategies used to promote foods to children. By "creative strategies" we mean the central creative propositions of advertisements. Many different appeals were identified, including appeals based on taste, nutritional/health properties, and fantasy/adventure. The most popular appeals used in the promotion of foods to children were hedonistic, including taste, humor, action-adventure, and fun. The use of animation techniques was found to be particularly strongly associated with children's food advertisements. Fast-food advertising, which has become more prominent in recent years, tends not to describe the product advertised but to focus on the experience of the meal and the brand.

The Effects of Food Promotion to Children

Is There a Causal Link Between Food Promotion and Children's Food Knowledge, Attitudes, and Behavior?

Thirty-three studies were judged to be capable of examining potential causal links between food promotion and children's food-related knowledge, preferences, and/or behavior. These were primarily experimental and cross-sectional studies (there was also one observational study and one quasi-experiment) that utilized methods and analysis procedures capable of providing evidence of a potentially *causal* relationship between food promotion and effects on children. Studies were assessed according to the Bradford-Hill (1965) criteria for inferring causality: temporality, reversibility, dose response, consistency, specificity, and plausibility. For more details of this process and the applied criteria, see the full review (Hastings et al., 2003).

The studies investigated a range of effects, including nutritional knowledge, food preferences, food purchase-related behavior, food consumption behavior, and diet.

Nutritional Knowledge

The studies that investigated the influence of food promotion on children's nutritional knowledge provided modest evidence of an effect. Four studies found that food promotion had an effect on or was associated with differences in nutritional knowledge: in three (Ross et al., 1980; 1981; Wiman and Newman, 1989; Gracey et al., 1996) exposure to food promotion for "low nutrition" foods was associated with poorer nutritional knowledge, and in one (Peterson et al., 1984) exposure to advertisements for foods "high in nutritional value" increased nutritional knowledge.

Three studies found that exposure to food promotion had no impact on, or was not associated with, changes in, children's nutritional knowledge (Atkin, 1975; Goldberg et al., 1978a; 1978b), and the eighth study (Galst, 1980) produced inconclusive results.

The evidence is modest rather than strong. Two of the studies (Ross et al., 1980; 1981; Gracey et al., 1996) were methodologically weak, and in another two (Galst, 1980; Peterson et al., 1984) it was difficult to separate out the effects of advertising from other exposure variables. Furthermore, studies that found effects tended to take more detailed knowledge measures than did the studies that did not find effects: the studies were not measuring the same effects.

Food Preferences

Thirteen randomized controlled experiments and one cross-sectional study investigated the influence of food promotion on children's food preferences; twelve reported relevant results. Of these, seven found that exposure to food promotion had an impact on, or was associated with, significant changes in children's food preferences (Goldberg et al., 1978a; 1978b; Gorn and Goldberg, 1980a; Heslop and Ryans, 1980; Stoneman and Brody, 1981; Kaufman and Sandman, 1983; Norton et al., 2000; Borzekowski and Robinson, 2001). Four of these were good quality experimental studies. Three of the four (Goldberg et al., 1978a; 1978b; Stoneman and Brody, 1981; Kaufman and Sandman, 1983) found that children were significantly more likely to prefer high-fat, -salt, or -sugar foods over lower fat, lower salt, or sugar alternatives after exposure to food advertisements; the fourth (Borzekowski and Robinson, 2001) found that children preferred the advertised brand over a nonadvertised brand of the same product type after food advertising exposure.

One study found nonsignificant effects (Goldberg et al., 1978a; 1978b), and four (three experiments and one cross-sectional study) found none (Ritchey and Olson, 1983; Peterson et al., 1984; Clarke, 1984; Gorn and Florsheim, 1985). These were generally weaker studies, in terms of design and relevance of measures and because the effects of food promotion could not be separated from other experimental stimuli.

Purchasing and Purchase-Related Behavior

Seven studies looked for these effects: three were randomized controlled experiments, one was a natural quasi-experiment, one observational study involved observations in a natural situation, and two were cross-sectional surveys.

All seven studies found that exposure to food promotion influenced, or was significantly associated with, purchase or purchase-related behavior. One experiment (French et al., 2001) found that promotional signage on vending machines significantly increased sales of low-fat snacks in secondary schools independently of pricing variables. The natural quasi-experiment (Goldberg, 1990) compared household cereal purchases of English- and French-speaking children in Montreal. The former watched primarily U.S. television, which carried children's food advertising; the latter watched primarily Quebec television, which had banned such advertising in 1980. Regression analysis showed that watching U.S. television was significantly related to increased purchase of advertised cereals independently of income or language.

The two remaining experimental studies found that exposure to food promotion significantly increased children's purchase-influence behavior observed while supermarket shopping with parents (Galst and White, 1976; Stoneman and Brody, 1982), and one observational and one cross-sectional study (Atkin, 1975; Reeves and Atkin, 1979) found significant associations between amount of Saturday morning television viewed and frequency of food purchase requests to parents, with "heavy" viewers in both studies making more requests than "light" viewers. The remaining study (Taras et al., 1989) found a weak association between television watching in general and food purchase requests to mothers.

Food Consumption

Eleven studies investigated the effects of exposure to food promotion on children's food consumption behavior. Five of the studies provided evidence of an effect on consumption behavior, and six did not.

Two experimental studies found significant effects on children's consumption behavior: in one, exposure to food promotion reduced the children's likelihood of selecting fruit or orange juice, compared with a sweet, for a daily snack (Gorn and Goldberg, 1982; Gorn and Goldberg, 1980b), and in the other it increased boys' calorific consumption from a tray of snack foods (Fox, 1981; Jeffrey et al., 1982). Three cross-sectional studies (Atkin, 1975; Ritchey and Olson, 1983; Bolton, 1983) found small associations, of varying degrees of strength, between exposure to television food advertising (as measured by television viewing) and frequency of snacking or consumption of specific foods.

Six studies reported inconclusive results. In two studies (Galst, 1980; Peterson et al., 1984), it was not possible to disentangle the effects of food promotion from other experimental stimuli. Two studies (Gorn and Goldberg, 1980a; Cantor, 1981) found that exposure to food promotion influenced consumption behavior (in divergent directions) under some conditions but not others. The last two studies (Jeffrey et al., 1982; Dawson et al., 1988) found nonsignificant effects.

Diet and Health

Four cross-sectional studies investigated the relationship between television viewing and children's diet (Bolton, 1983; Taras et al., 1989; Gracey et al., 1996; Coon et al., 2001), and two explored the relationship between television viewing and health-related variables: obesity (Dietz and Gortmaker, 1985) and cholesterol levels (Wong et al., 1992). The former study had a longitudinal element.

All four diet studies found significant associations, of varying strength, between television viewing and children's calorific intake and nutrient efficiency (i.e., the proportion of nutrient requirements satisfied to energy [calorific] requirements satisfied) (Bolton, 1983), intake of meat, salty snacks, and soda (Coon et al., 2001), fat consumption (Gracey et al., 1996), and calorific intake (Taras et al., 1989). Of the two studies examining health-related variables, one (Wong et al., 1992) found that time spent watching television and playing video games was a significant and independent predictor of raised cholesterol in children, whereas the other (Dietz and Gortmaker, 1985) found that television viewing was predictive, at marginally significant levels, of obesity and obesity in 3 to 4 years time and that this effect occurred independently of prior obesity and family socioeconomic characteristics.

In five of the studies (Dietz and Gortmaker, 1985; Taras et al., 1989; Gracey et al., 1996; Wong et al., 1992; Coon et al., 2001), the potential effect of food advertising could not be disentangled from that of television viewing. Thus, the associations found may be attributable to the impact of food advertising seen while watching television, other television messages (e.g., programming), the sedentary nature of television viewing, or some other variable. One study, however (Bolton, 1983), used detailed viewing diaries, which enabled a calculation of the amount of food advertising to which each subject was exposed. A structural equation model found that the greater a child's food advertising exposure, the more frequent his or her snacking and the lower his or her nutrient efficiency.

If Food Promotion is Shown to Have an Effect on Children's Food Knowledge, Preferences and Behavior, what is the Extent of this Influence Relative to other Factors?

Eight studies examined the influence of food promotion or television viewing compared with one or more other factors (such as age, race, gender, parental snacking, and parental attitudes): seven cross-sectional studies (Ritchey and Olson, 1983; Bolton, 1983; Dietz and Gortmaker, 1985; Wong et al., 1992; Gracey et al., 1996; Norton et al., 2000; Coon et al., 2001) and one experiment (French et al., 2001).

One study (Bolton, 1983) found that food advertising exposure had a small but significant impact on children's snacking frequency, nutrient efficiency, and, indirectly, calorie intake. The effect occurred independently of parental snacking frequency, child's age, parental diet supervision, and missed meals, but food advertising exposure explained less of the variance than parents' snacking frequency. One study (Ritchey and Olson, 1983) compared the influence of television watching on children's consumption of sweets with the influence of parents' frequency of consumption of sweet foods and parents' attitudes toward sweet foods. Parents' frequency of consumption had a clearer impact than television watching, but television watching had a stronger impact than parents' attitudes.

The one experimental study (French et al., 2001) found that promotional signage (highlighting or encouraging low-fat snack options) on vending machines significantly increased low-fat snack sales in secondary schools independently of a range of different pricing strategies.

The remaining studies (Dietz and Gortmaker, 1985; Wong et al., 1992; Gracey et al., 1996; Norton et al., 2000) provided evidence, of varying strength, that food promotion and television viewing have a significant influence on children's food behavior and diet independent of at least one other factor but did not examine or report the relative strength of the different influences.

In the Studies that Demonstrate an Effect of Food Promotion on Children's Food Knowledge, Preferences, and Behavior, does this Effect Impact the Total Category Sales, Brand Switching, or Both?

Only two studies (Gorn and Goldberg, 1980a; Gorn and Florsheim, 1985) examined both brand and category effects, but these were of limited value: one omitted details of the product categories and the other dealt with adult food advertising. However, 13 studies did provide good or reasonable quality data on brand and category effects individually. Of the five studies that examined whether food promotion influenced brand preferences, two (Gorn and Goldberg, 1980a; Borzekowski and Robinson, 2001) found that food promotion encouraged children to prefer the advertised over a non-advertised brand, two found that it had no impact (Clarke, 1984; Gorn and Florsheim, 1985), and one found only very modest impacts in favor of the advertised brand (Heslop and Ryans, 1980).

Of the eight studies that compared children's preferences or behavior in relation to foods in different categories, four found that they were more likely to select higher fat, sugar, or salt products (compared with lower fat, sugar, or salt alternatives) in a one-off preference test (Goldberg et al., 1978a; 1978b; Stoneman and Brody, 1981; Kaufman and Sandman, 1983) or for a daily snack (Gorn and Goldberg, 1980b; 1982). The fifth study (Goldberg et al., 1978a; 1978b) found no significant impacts on category preferences. Three of the studies produced inconclusive results (Galst, 1980; Cantor, 1981; Peterson et al., 1984).

CONCLUSIONS

Children's food promotion is dominated by television advertising, including advertisements for presugared breakfast cereals, confectionary, savory snacks, soft drinks, and fast-food outlets. The advertised diet varies greatly from the recommended one, and themes of fun and fantasy or taste, rather than health and nutrition, are used to promote this to children.

The review addressed the central question of whether promotional activity actually has an impact on children. This task was not straightforward. Food knowledge, preferences, and behavior are influenced by a wide range of complex and dynamic factors. Unpicking these is difficult, and isolating the possible influence of just one variable—in this case promotion—particularly so. Moreover, social science research of this ilk can never provide final incontrovertible proof. It reduces uncertainty rather than produces certainty and proceeds on the basis of testing plausible hypotheses and making judgments on the balance of probabilities. The review had to identify all of the relevant studies, assess their quality, and reach a composite judgment on what the literature can tell us about the problem.

Despite gaps in the evidence base (e.g., many studies examined only television advertising, very few studies measured the strength of promotional impacts *relative* to other factors influencing children's food choices), the review demonstrates that there is a link between advertising and children's food knowledge, preferences, behavior, and diet/health status. Furthermore, it is clear from a number of good-quality studies that these effects are independent of other influences, such as parental behavior and price.

It is also clear that effects occur at both a *brand* and a *category* level. In other words, advertising can shift children's preferences not just between different brands of chocolate cookies, but between chocolate cookies and fruit. Many people are prepared to accept brand effects but not category ones. The review shows that there is robust evidence that food advertising achieves both kinds of effects.

IMPLICATIONS FOR OBESITY PREVENTION

Although the review did not directly examine the link between food promotion and *obesity*, its findings have important implications for the dietary health of children.

First, the review demonstrates that the majority of children's advertisements are for products high in sugar, fat, and salt. Second, it provides evidence that this kind of promotional activity *is* affecting children, particularly in terms of the food they express a preference for, buy, and ask their parents to buy. Given the nature and extent of food promotional activity (and its ability to influence young people), it would be reasonable to conclude that the current promotional climate is encouraging children to make unhealthy rather than healthy choices. It is also likely that this will be having an impact on their dietary health.

We cannot reasonably expect children to make healthy choices in an environment that aggressively promotes the opposite (Bronwell and Battle-Horgen, 2003). Currently, messages from the public health sector struggle to compete with the vast efforts and resources of the food industry. Parents are often encouraged to exert more control over their children's dietary choices, but their efforts are hindered by the heavy advertising of less healthy foods.

The government is currently deciding how best to tackle the problem. Action is needed to "detoxify" the current environment and make it easier for both children and parents to make healthier choices. One very likely possibility involves introducing targets to increase promotional activity for healthier foods, i.e., foods that are lower in salt, sugar, and fat. Encouragingly, the review found some evidence that food promotion could also influence children to make healthier choices. It seems perfectly possible for healthier food products to be promoted to children using the same techniques and strategies currently used to sell sugary drinks and salty snacks. The food industry clearly has the power to influence dietary choice: it is the type and mix of food that they promote that will determine whether their influence is harmful or beneficial.

Finally, policy action on food promotion should form part of a broader strategy to tackle obesity among young people. Effectively addressing the problem of obesity will require a comprehensive, coordinated and preferably government-initiated strategy that involves action on all policy fronts, not just advertising.

ACKNOWLEDGMENTS

We acknowledge the contribution of our colleagues who collaborated on the project. They are Dr. Alasdair Forsyth (Centre for Social Marketing), Anne Marie Mackintosh (Centre for Social Marketing), Dr. Mike Rayner (Oxford University), Professor Christine Godfrey (University of York), Dr. Martin Caraher (London City University), and Kathryn Angus (Centre for Social Marketing). This research was funded by the Food Standards Agency.

REFERENCES

Atkin, C.K., *The Effects of Television Advertising on Children, Report No. 6: Survey of Pre-Adolescent's Responses to Television Commercials,* Office of Child Development (DHEW), Washington, DC, 1975.

Bolton, R.N., Modeling the impact of television food advertising on children's diets, in *Current Issues and Research in Advertising,* Leigh, J.H. and Martin, Jr, C.R., Eds.,

Division of Research, Graduate School of Business Administration, University of Michigan, Ann Arbor, 1983, pp. 173–199.

Borzekowski, D.L.G. and Robinson, T.N., The 30-second effect: an experiment revealing the impact of television commercials on food preferences of preschoolers, *J. Am. Diet. Assoc.*, 101, 42–46, 2001.

Bradford Hill, The environment and disease: association or causation, *Proc. R. Soc. Med.*, 58, 295, 1965.

Bronwell, K. and Battle-Horgen, K., *Food Fight: The Inside Story of the Food Industry, America's Obesity Crisis, and What We Can Do About It*, 1st ed., McGraw Hill, New York, 2003.

Cantor, J., Modifying children's eating habits through television ads: effects of humorous appeals in a field setting, *J. Broadcasting*, 25, 1, 37, 1981.

Clarke, T.K., Situational factors affecting preschoolers' responses to advertising, *Academy Market. Sci. J.*, 12, 4, 25, 1984.

Coon, K.A., Goldberg, J., Rogers, B. L., et al., Relationships between use of television during meals and children's food consumption patterns, *Pediatrics*, 107, 1, e7, 2001 (electronic article), available at http://pediatrics.aappublications.org.

Crockett, S. and Sims, L., Environmental influences on children's eating, *J. Nutr. Educ.*, 27, 235, 1995.

Dawson, B.L., Jeffrey, D., Balfour, et al., Television food commercials' effect on children's resistance to temptation, *J. App. Soc. Psyc.*, 18, 16, 1353, 1988.

Dietz, W.H. and Gortmaker, S.L., Do we fatten our children at the television set? Obesity and television viewing in children and adolescents, *Pediatrics*, 75, 5, 807, 1985.

Escobar, A., Factors influencing children's dietary practices: a review, *Family Econ. Nutr. Rev.*, 12, 45, 1999.

Fox, D.T., *Children's Television Commercials and Their Nutrition Knowledge and Eating Habits*, PhD. Thesis, University of Montana, MT, 1981.

French, S.A., Jeffery, R.W., Story, M., et al., Pricing and promotion effects on low-fat vending snack purchases: The CHIPS Study, *Am. J. Public Health*, 91, 112–117, 2001.

Galst, J.P. and White, M.A., The unhealthy persuader: the reinforcing value of television and children's purchase influencing attempts at the supermarket, *Child Dev.*, 47, 1089, 1976.

Galst, J.P., Television food commercials and pro-nutritional public service announcements as determinants of young children's snack choices, *Child Dev.*, 51, 3, 935, 1980.

Goldberg, M.E., A quasi-experimental assessing the effectiveness of TV advertising directed to children, *J. Market. Res.*, 27, 4, 445, 1990.

Goldberg, M.E., Gorn, G.J., and Gibson, W., The effects of TV messages for high and low nutritional foods on children's snack and breakfast food choices, *Adv. Cons. Res.*, 5, 540, 1978a.

Goldberg, M.E., Gorn, G.J., and Gibson, W., TV messages for snacks and breakfast foods: do they influence children's preferences?, *J. Cons. Res.*, 5, 2, 73, 1978b.

Gorn, G.J. and Florsheim, R., The effects of commercials for adult products on children, *J. Cons. Res.*, 11, 4, 962, 1985.

Gorn, G.J. and Goldberg, M.E., Children's responses to repetitive television commercials, *J. Cons. Res.*, 6, 4, 421, 1980a.

Gorn, G.J. and Goldberg, M.E., *TV's Influence on Children: The Long and the Short of It*, presented at the 88th Annual Convention of the American Psychological Association, Montreal, Quebec, Canada, September 1–5, 1980, 1980b.

Gorn, G.J. and Goldberg, M.E., Behavioral evidence of the effects of televised food messages on children, *J. Cons. Res.*, 9, 2, 200, 1982.

Gracey, D., Stanley, N., Burke, V., et al., Nutritional knowledge, beliefs and behaviours in teenage school students, *Health Educ. Res.*, 11, 2, 187, 1996.

Hastings, G., Stead, M., McDermott, L., et al., *Review of Research on the Effects of Food Promotion to Children—Final Report, Report to the Food Standards Agency*, University of Strathclyde, Centre for Social Marketing, Glasgow, 2003.

Heslop, L.A. and Ryans, A.B., A second look at children and the advertising of premiums, *J. Cons. Res.*, 6, 4, 414, 1980.

Jeffrey, D.B., McLellarn, R.W., and Fox, D.T., The development of children's eating habits: The role of television commercials, *Health Educ. Q.*, 9, 174–189, 1982.

Kaufman, L. and Sandman, P.M., *Countering Children's Sugared Food Commercials: Do Rebuttals Help?*, presented at the 34th Annual Meeting of the International Communication Association, San Francisco, CA, May 24–28, 1983.

Norton, P.A., Falciglia, G.A. and Ricketts, C., Motivational determinants of food preferences in adolescents and pre-adolescents, *Ecol. Food Nutr.*, 39, 3, 169, 2000.

Peterson, P.E., et al., *Do Pro-Nutritional Television Messages Improve Children's Eating Behaviour?: Empirical Findings and Recommendations for Further Research*, presented at the Annual Meeting of the American Psychological Association, Los Angeles, CA, August, 1984.

Reeves, B. and Atkin, C.K., *The Effects of Televised Advertising on Mother-Child Interactions at the Grocery Store*, presented at the 62nd Annual Meeting of the Association for Education in Journalism, Houston, TX, August 5–8, 1979.

Ritchey, N. and Olson, C., Relationships between family variables and children's preference for consumption of sweet foods, *Ecol. Food Nutr.*, 13, 257, 1983.

Ross, R.P., et al., *Children's Television Commercials Containing Nutritional Information: When Do They help? When Do They Hinder?*, presented at the Biennial Meeting of the Southwestern Society for Research in Human Development, Lawrence KS, March 27–29, 1980.

Ross, R.P., Campbell, T., Huston-Stein, A., et al., Nutritional misinformation of children: A developmental and experimental analysis of the effects of televised food commercials, *J. Appl. Dev. Psyc.*, 1, 4, 329, 1981.

Stoneman, Z. and Brody, G.H., Peers as mediators of television food advertisements aimed at children, *Dev. Psyc.*, 17, 6, 853, 1981.

Stoneman, Z. and Brody, G.H., The indirect impact of child-oriented advertisements on mother-child interactions, *J. App. Dev. Psyc.*, 2, 369, 1982.

Taras, H., et al., Television's influence on children's diets and physical activity, *J. Dev. Behav. Paediatr.*, 10, 4, 176, 1989.

WHO/FAO. *Diet, Nutrition and the Prevention of Chronic Diseases. Report of a Joint Food and Agriculture Organization/World Health Organization Expert Consultation*, WHO Technical Report Series 916, World Health Organization, WHO, Geneva, 2003.

Wiman, A.R. and Newman, L.M., Television advertising exposure and children's nutritional awareness, *J. Acad. Marketing Sci.*, 17, 2, 179, 1989.

Wong, N.D., Hei, T. K., Oaqundah, P. Y., et al., Television viewing and pediatric hypercholesterolemia, *Pediatrics*, 90, 75, 1992.

INDEX

Page number followed by t indicates table(s), f indicates figure(s)